A Pocket Guide to Poultry Health and D.

FIRST EDITION

Paul McMullin

5M Enterprises Limited

Copyright © Paul McMullin 2004
Published by 5M Enterprises Ltd.
PO Box 233
Sheffield
S35 0PB
United Kingdom

All rights reserved. No reproduction, copy or transmission of this publication may be made without written permission of the copyright holder.

ISBN 0-9530150-5-X

Printed and published in the United Kingdom, January 2004.

No part of this publication may be reproduced, copied or transmitted save with written permission or in accordance with the provisions of the Copyright Act 1956 (as amended), or under the terms of any licence permitting copying issued by the Copyright Licensing Agency, 90 Tottenham Court Rd, London W1T 4LP, Tel 020 7631 5555, Fax 020 7631 5500, email cla@cla.co.uk, web www.cla.co.uk

Any person who does any unauthorised act in relation to this publication may be liable to criminal prosecution and civil claims for damages.

A CIP catalogue record for this book is available from the British Library.

5M Enterprises Limited would welcome your comments regarding this book. All feedback will be taken into account and hopefully help enhance subsequent editions.

Please contact us at:

>5M Enterprises Limited
>PO Box 233
>Sheffield, S35 0AF
>United Kingdom
>Tel/Fax: 00 44 114 286 4638
>Email: info@thepoultrysite.com
>Web: www.ThePoultrySite.com

Updates of the information in this book can be found on our website at www.thepoultysite.com

ABOUT THE AUTHOR

Paul F. McMullin MVB, MRCVS, MSc, DPMP.

Paul McMullin is the senior veterinarian at Poultry Health Services, a specialist poultry practice in the UK. He is a past Honorary Secretary of the British Veterinary Poultry Association, a Member of the Royal College of Veterinary Surgeons, a diplomate and RCVS recognised specialist in Poultry Medicine and Production, and has recently completed an MSc in Livestock Health & Production from the University of London For the past 25 years he has provided consultancy services to a number of breeding and large and small commercial operations both in Europe and South America. He has published over 40 papers on management and disease control and lectured extensively to many farming and veterinary groups.

Poultry Diseases Pocket Guide

Note to the Reader

The methods of prevention, treatment and control of conditions discussed in the book are <u>guidelines only</u>. Any recommendations given and so used are the responsibility of the producer, and the advice of his or her veterinarian should be sought in case of doubt. No responsibility is accepted by the authors or publishers for any application of the advice given in this book because each farm and region is different and responses cannot be predicted. Some trade names and their chemical compounds are used throughout. No endorsement is intended nor is any criticism implied of similar products not named.

Acknowledgements

This book was inspired by the companion volume "Managing Pig Health and the Treatment of Disease" written by Mike Muirhead and edited by Tom Alexander.

Turning the inspiration into the reality involved quite a few people: The team at 5M who initially proposed this project and provided helpful comments on the drafts.

My particular thanks are due to Kate Turner Haig MRCVS who was instrumental in organising the information I provided into the first draft of this volume and also carried out the final checks. Both Kate and my colleague Kenton Hazel MRCVS also provided helpful comments on the manuscript. A number of other colleagues kindly reviewed and improved the manuscript: Bob Henry MRCVS who reviewed the section on diseases of ducks and geese and provided many helpful comments on the relevance of other diseases to these species, Perpetua McNamee MRCVS, who reviewed bacterial diseases, and Rob Davies MRCVS who reviewed the main zoonoses.

Although this book is clearly not an atlas of disease, from the outset we wished to include illustrations to give an idea of the range of disease effects seen in practice. Our thanks to Bob Henry for Figures 8 and 40, to Jannsen Animal Health for Figure 11, to Bayer Animal Health for figures 12 to 16, to Ken Kirkpatrick MRCVS for Figure 26, and to Dr Jim Trites DVM of Elanco Animal Health for figures 37 and 38.

Acknowledgements

In summary, many people have contributed to bringing you what we hope is a useful handbook of poultry health and disease. However the responsibility for any errors that may have crept in rests entirely with me.

As Mike also wisely noted in his acknowledgements: "No book is ever complete (or without fault) but we hope that this one will provide scope for debate, education, improvements in health and disease control and …. more profitable …. farming."

Finally and most especially to my wife Maria, and sons Christian and Nicholas, not just for their constant support but also for their patience in waiting for me to "emerge from behind the laptop" during projects such as this!

Paul McMullin
Thirsk
January 2004

CONTENTS

Chapter 1 - Introduction	3
Satisfying market demands	4
A model of health and disease in poultry populations	5
Managing for poultry health	5
Chapter 2 - Principles of biosecurity	9
Chapter 3 - The immune system: vaccines and how they work	29
Chapter 4 - Poultry medication – a practical guide to medicines and feed additives	35
Legal Requirements	38
How Medicines are Prescribed	40
Understanding Dosage Levels	40
Administering Medicines	41
Administering Medicines Orally	41
Other Types of Medicine for use on the Farm	45
Veterinary Health Plans and Planning	56
Chapter 5. Diseases and syndromes: Introduction	61
Occurrence of disease by poultry species - Index	65
Chapter 6. Diseases and syndromes: Chickens and Various Species	71
Chapter 7. Diseases and Syndromes: Predominantly of Turkeys.	213
Chapter 8. Diseases and Syndromes: Predominantly of Ducks and Geese.	239
Appendix 1. A Glossary of Technical Terms, and Abbreviations	252
Appendix 2 – One answer to the Task A	259
Appendix 3 - Some Useful Tables	261
Index	271

INTRODUCTION

CHAPTER 1

INTRODUCTION

Poultry production has grown enormously in most countries over the past 50 years. Whether kept as small-scale flocks or in large commercial enterprises, through eggs or meat, poultry now contribute a significant part of the high quality animal protein consumed by the human population. The market for poultry products is still growing substantially in many countries, and is highly competitive.

Today chickens, turkeys and ducks, are farmed in most countries. The phenomenal growth in production of poultry meat in the latter half of the 20th century has been made possible by a number of developments:

- Genetic improvement
- Artificial incubation on an industrial scale
- Control of infectious and parasitic disease (allowing the keeping of larger individual flocks, farm and area populations) achieved through a number of mechanisms:
    o routine antiparasitic medication (e.g. for coccidiosis)
    o control of endemic viral infections by vaccines (e.g. for Marek's Disease)
    o control of epidemic viral infections by vaccines and isolation (e.g. Newcastle Disease)
    o eradication of important bacterial infections by separation of the generations, selective medication, and isolation of stock (*Mycoplasma* and some *Salmonella* infections)
    o improved nutrition.

Success in modern poultry production depends to a very considerable extent on the ability of the managers and stock handlers to appreciate and respond to the needs of the birds. Even where the scale of production is industrial, the underlying biological and agricultural processes involved remain very important. Increasing scale and mechanisation can increase the impact, for good, or for bad, of the individual manager or stock handler on the well-being of the birds.

Satisfying market demands

Market requirements are already having a significant effect on how we control diseases in poultry. This effect is likely to increase from here on. Health-related market requirements relate to the control of foodborne infections capable of causing significant disease in man, avoidance of residues, and use of antibiotics. The recently announced significant improvement in the Salmonella status of UK broiler meat production is certainly a reflection of all-in/all-out production and better understanding of the sources of the infection. There are demands to know 'what chemicals are used' and to control residues. The UK, for example, has one of the most sophisticated residues testing systems in the world and regularly publishes its results. It concentrates on liver residues because these are a more sensitive indicator than muscle samples. The levels of residues in eggs have improved substantially, and nicarbazin residues in broiler livers have improved considerably. A working group has published guidelines on the avoidance of nicarbazin residues (see page 54).

Notifiable diseases (currently Newcastle Disease and high pathogenicity Avian Influenza) would not previously have been considered an issue of interest to consumers. However, the serious losses from Avian Influenza in Italy in 1999–2000, the Netherlands in 2003, and experience with Swine Fever and Foot & Mouth Disease in the UK may well affect our ability to rapidly and effectively deal with these conditions in the future. Governments are increasingly concerned about the cost, publicity about 'stamping out' policies and knock-on effects on other industries such as tourism.

At Poultry Health Services our purpose-designed computer systems allow us to collect and analyse data that help our clients satisfy market demands in a number of areas. They also help us create and document health plans that are a good starting point for a disease control programme. Most assurance schemes now require such health plans though they do not usually specify how detailed they need be.

Chapter 1 - Introduction

This handbook is the first in a series aimed at poultry farmers, and fieldsmen. Agriculture and veterinary students and veterinarians dealing with small poultry flocks should also find it a useful reference. It is intended to provide background information to help poultry farmers make better use of their veterinarian and is not a substitute for the use of the veterinarian. In an increasing number of countries veterinarians are required, either by legislation, codes of practice or production standards agreed with major customers, to be directly involved in poultry-health decision making and medication.

A model of health and disease in poultry populations

Poultry production systems, like all other farming activities and regardless of scale, are biological systems in which living organisms are managed to achieve a particular goal. Poultry medicine concentrates heavily on preventing disease, and, where treatment is required, medication of whole flocks of birds. In order to understand why disease occurs and how to set about avoiding it in future, we need a thorough understanding of the production system and of poultry management.

Managing for poultry health

What is management? Figure 1 below shows one way of looking at a model poultry production system. We could take all of the known and unknown requirements of the particular bird (species, strain, age, sex) as determined by its genotype and line them up in a stack on the left of the figure. This would provide a 'chart' of its requirements. We could then, for each of these requirements, line up what the natural environment provides on the right of the chart. A gap would be left in the middle. Management is, essentially, everything that we do that tends to 'fill the gap'. It will have a broad range of components: temperature control (heating and, potentially, cooling), ventilation, lighting, formation of groups, feed composition, feed distribution, feeder design, drinker design, and so on. Some of these will be identical or very similar among different classes of poultry; some may require subtle variations even between different strains. The type, intensity and efficiency of management will, inevitably, vary. Factors that will have an impact are basic economics (labour cost, capital cost, product value), climate, season, availability of labour, experience of stockpeople, and production system (e.g. intensive or free-range). Management, at least as conventionally defined, is never perfect. One way of looking at the deficits is to call them 'health and welfare black holes'. Whole flocks, farms, even companies or breeds may disappear down them if we do not take care with these!

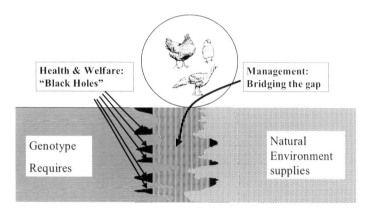

Figure 1 – Bridging the Gap

Another way of looking at the gaps is to call them 'stress'. Stress is the physiological adaptation mechanism that comes into play to help the bird adapt to its environment.

So, what do we do about trying to fill as many of these gaps as well as possible? We are, in fact filling them all the time, even when we don't realise it. Every time we use a vaccine, a disinfectant, a feed additive or a medicine we are helping the stock adapt to a particular challenge or deficit between what it needs and what management can provide (figure 2). In fact, all of these measures could be regarded, in the broadest sense, as components of management, by the definition we gave above.

Breed manuals are a very good starting point for the development of optimal management systems. However we need to keep in mind that they are, of necessity, very general – the environments in which birds are kept may require modifications of these general recommendations. Primary breeding companies are constantly evaluating how their birds are performing under practical farming conditions. They will usually have access to test facilities in which their products are managed under standard conditions. Their objectives in this are to provide feedback to the selection programme and to better inform their customers about what works best. However they probably will not have direct in-house experience of managing the product bird in all of the broad range of farming environments in which it is grown.

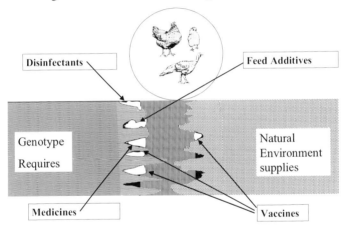

Figure 2 – Supplementing management

We need to keep in mind that the areas in the figures above are not static. Genotypes change over time. Usually the effects of this are subtle and very gradual. Sometimes the change can appear to occur quite suddenly. The natural environment also varies between night and day, with season, and perhaps over longer periods (e.g. long-term climate change or global warming).

PRINCIPLES OF BIOSECURITY

CHAPTER 2

PRINCIPLES OF BIOSECURITY

Principle 1.

Everyone involved in poultry production, whether owner, manager, farm worker, contractor, driver, or veterinarian, must have a sound knowledge of the objective of biosecurity and what it means in practice.

Biosecurity – one definition:

'All of the measures which can or should be taken to prevent viruses, bacteria, fungi, protozoa, parasites, insects, rodents, and wild birds from entering or surviving, and infecting or endangering the wellbeing of the poultry flock.' (K.Gooderham)

Biosecurity is almost always a good thing, even though, as we will see below, it sometimes has unintended consequences.

The fundamental principle involved in biosecurity is the avoidance of contact with a source of disease or contagion. The contagious nature of disease has been evident to man for hundreds if not thousands of years even though the precise cause was not usually evident. While many insects, as well as rodents and wild birds, are easily seen, the threat that they represent is at least partially due to the microscopic organisms they carry. These only became visible with the advent of good microscopes – Antonie van Leeuwenhoek described 'animalcules' in intestinal contents and teeth scrapings in the 1670s. However it was almost 200 years before Louis Pasteur and Robert Koch demonstrated the connection between microbes and disease. In 1881 Pasteur described *Pasteurella multocida*, the cause of Fowl Cholera, and showed how the disease could be prevented.

Poultry Diseases Pocket Guide

There are three characteristics of microbes that make biosecurity both necessary and difficult to implement: their small size, the often long interval between exposure and an observable effect in the birds, and the ability of organisms to survive in the environment. Periods of survival outside the host vary markedly between different organisms – from less than an hour (*Histomonas*) to many years (spores of *Clostridium*).

In discussing size of organisms the following table may be helpful:

Organism	Size (Microns)	Number per cm
Salmonella spp.	1 to 3	3333
Pasteurella spp.	0.6 to 2.5	4000
Mycoplasma gallisepticum	0.25 to 0.5	20 000
Infectious Bronchitis virus	0.08 to 0.1	100 000

Organisms vary in size according to species and there is also a moderate amount of natural variation within species. The other reason for the range of size shown above is that some organisms (e.g. Salmonella) are oblong, so they have one short and one long axis. One way of conveying the very small size of these organisms is to show how many would need to be lined up lengthwise in a centimetre, as shown in the column on the right. To put it another way, if a salmonella cell were 3 cm long, a chicken would be about 3 km high!

One consequence of the small size of microbes is that a very larger number may be present in a small amount of material. A single pellet of mouse faeces can contain 230 000 *Salmonella* Enteritidis organisms. This is enough to infect up to 5000 chickens. Viral particles are often present in even larger numbers. It has, for instance, been calculated that 1 gm of chicken manure can contain enough viral particles to infect 1 million birds with avian influenza.

This information can usefully be incorporated into staff training packages on biosecurity. Such training should be included on staff induction and on a regular basis. Where possible this training should be adapted to the local conditions and hazards. A log of who has received which training, and when, should be maintained.

Principle 2.

The immediate environment of the stock is considered to be 'clean' and everything that is outside the environment is considered to be 'dirty'. Anything that moves from the 'dirty' area to the 'clean' area should be subject to control measures appropriate to likely risk and status of the stock being protected.

This relates to the microscopic size of infectious organisms and the second characteristic of microbial infection mentioned above – infected birds may appear normal. This leads to the assumption that all other flocks are a potential source of infection, as is the external environment.

This principle is implemented most economically if it is incorporated in the initial planning of new enterprises or refurbishments of existing ones. It will be necessary to balance the benefits brought by biosecurity against the initial costs of the required installations and the extra ongoing costs of operating the systems. Ongoing costs will include such things as overalls and protective clothing (disposable or washable), movement restrictions (and their implications on staff time and efficiency), and disinfectants.

In setting an appropriate level of biosecurity for an operation it is normal to consider at least the following aspects:

- The type (e.g. pure line, grandparent, parent or commercial) and individual value of the stock.
- The number of birds per flock, per site and in the immediate area.
- Any official disease control programmes affecting the business.
- Known or estimated infection status of other sites with operational links to the particular facility (farms, hatcheries, feed mills).
- Known or estimated infection status of other poultry farms in the immediate area.

- Presence of other farm animals on site and their estimated infection status. This may be extended to other farm animals in the immediate area.
- Presence of wildlife in immediate farm environment, especially migratory birds.
- Any legally binding or customer requirements with respect to the infection status of product originating on the site.

Figure 3. Biosecurity – Shielding the Pyramid

It is recognised that the level of biosecurity applicable in a production system will need to be tailored to the particular circumstances. The larger the numbers of birds, the more valuable the stock, and the greater the risks of infectious disease, the higher the biosecurity barrier that may be justified. In general terms, the biosecurity shield will tend to be less robust as we move down the production pyramid, as illustrated in the figure 3 above.

The assessment of required biosecurity is basically a formal or informal analysis of the infectious risks to which the stock will be exposed and the severity of the consequences. It may be helpful to carry out economic analyses illustrative of a couple of disease scenarios. To be realistic such analyses would need to take some account of the indirect costs of a significant disease outbreak. Emergency slaughter of a breeding chicken flock, for instance, would result in major costs, and risks, in replacing the lost hatching eggs.

Chapter 2 - Principles of Biosecurity

If the company and/or farms operate an HACCP (Hazard Analysis and Critical Control Point) programme, the above risk assessment can be carried out as part of the programme and the resultant plans and procedures should then be documented in the HACCP plan. HACCP plans within poultry companies often originate in processing plants and their scope is restricted to biological and chemical risks relevant to the safety of the finished product as it affects humans. The same principles may be applied to biosecurity at farm level if we extend the scope to include all hazards to the health of the poultry. The maintenance of a full HACCP system may be considered too burdensome for an agricultural production operation but the general approach of identifying hazards, implementing control measures, and monitoring that they are working is a sound basis for an effective biosecurity system. Given increased interest in a 'farm-to-fork' approach to the control of hazards in the food chain HACCP is likely to be increasingly applied to agricultural production. Practical implementation of HACCP and HACCP-like systems are covered elsewhere in this series.

Major components of biosecurity include:

- Allowing only necessary visitors to production sites – most sites will have a 'quarantine period' applied to visitors with access to sites in other production systems, typically of 3–7 days. Compliance needs to be documented and monitored through the use of a visitors log at each site.
- Restricting movement of workers and equipment between houses, sites and age groups. Here too it may be necessary to implement quarantine.
- Providing sanitising foot baths, showers and protective clothing at strategic locations.
- Maintaining cleaning and disinfection programmes, especially in hatcheries.
- Reducing microbial load on vehicles and other mobile equipment by washing and disinfecting at critical times.
- Locating production sites strategically in relation to other production sites and movement of poultry, thus minimising transfer of disease.

Poultry Diseases Pocket Guide

- Restricting contact of workers with other poultry, especially potential carriers of hazardous disease organisms.
- Appropriately handling waste and dead birds to minimise the transfer of disease between sites.
- Controlling rodents and wild birds effectively, since both are potential disease vectors. Rodents, insects and wild birds can harbour many pathogens that will cause disease and infections in poultry. An integrated system of control of the numbers of these animals (where legally permitted) and limiting contact between them and poultry is advised. Monitoring systems should be used to ensure that action is taken in the early stages of a population rise rather than afterwards. They are particularly important during farm depopulation.

A simplified diagram, figure 4 of a high-biosecurity site design is shown below. Each of its components is worthy of some comments.

<u>Perimeter fence</u> – This should exclude easy access to people and domestic animals. If the site has free-range poultry it may also be helpful for it to be predator-proofed.

<u>Entrance block</u> – This will often house the site office, staff and visitor changing areas, toilets, showers. A communication system between the gate and the rest of the site is necessary so that visitors can announce themselves to farm staff easily and reliably (e.g. bell or pager). High-security sites will normally feature storage for street-clothes, walk-through showers, storage for clean clothes for use on site, and a fumigator to allow fumigation of small pieces of equipment or other materials.

<u>Service area</u> – This is the final control point before access to bird areas. It may incorporate showering, change of outer clothing, change of footwear, use of disposable overshoes, or foot dips in various combinations, depending on the level of biosecurity required. A well-maintained concrete pad in front of the house entrance will reduce the risks of contaminants being tracked into this area. If boot-changing is practised it helps to have a physical separation of the clean and dirty areas, for example with a step-over barrier or seat. If

Chapter 2 - Principles of Biosecurity

possible the separation of the two areas should be designed in such a way as to give access to control panels from either area. An inspection window to be able to view conditions in the house without disturbing the birds is also helpful.

Clean and dirty routes – Most sites operate with only a single entrance route. Where a number of houses are arranged in nuclei within a larger complex it may be advisable to arrange double routes to minimise the risks associated with multiage complexes. The 'dirty route' would then be used mainly for removal of manure/litter, dead birds and final stock depletion.

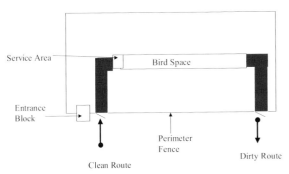

Figure 4 A Biosecurity System

Principle 3

Each site should have the smallest number of bird types and ages consistent with the business objectives.

The multistage nature of poultry production (grandparent flock, hatchery, parent flock, hatchery, commercial bird, product), combined with the very short 'shelf-life' of intermediate products (hatching eggs, chicks), means that production planning can have an enormous impact, both for good and for bad, on the health status of a production system. Wherever possible all-in/all-out production should be planned on a site basis. The time required for effective cleaning and disinfection between flocks will depend on many factors such as the type of equipment, surface finish, state of repair and so on. Whatever the planned 'turn-around', compliance with the plan should be monitored over time.

Principle 4

> *Appropriate decontamination procedures should be in place between one crop and the next.*

The normal 'dirty'/'clean' separation as outlined in Principle 2, can be extended to separation in time between one crop and the next. In this case we consider everything that had contact with the previous flock to be 'dirty' until such time as it is effectively decontaminated. Full clean-out and disinfection is the norm for all breeding birds and many laying and commercial meat birds in various countries. The period between one flock and the next is commonly referred to as 'downtime' or 'turn-around'. It introduces a daily cost which will vary with the amount of capital employed on the site. In countries where litter is not removed from commercial birds at every crop it is customary to process the litter and follow an extended turn-around compared to farms that have been washed and disinfected.

Cleaning and disinfecting houses and equipment

When poultry are removed from houses, the buildings and equipment should be carefully cleaned and disinfected before new birds are introduced. Manure (including litter) should be removed from the immediate vicinity of the poultry houses, preferably to an off-site location. A successful cleaning and disinfection protocol should cover:

- Plans to include site specific issues such as required maintenance.
- Removal of birds, checking rodent bait.
- Removal of mobile equipment.
- Removal of litter and as much other material as possible.
- Washing to remove maximum organic material.
- Cleaning/sanitising the water system.
- Cleaning/sanitising all surfaces – record concentration and usage.
- Cleaning/sanitising equipment.
- Set up of equipment.
- Fogging.

The appropriate detergents and disinfectants will vary with the nature of the production system and disease or infection challenge. In all cases, however, effective cleaning, and careful identification and separation of unsanitised and sanitised areas/materials will maximise the efficacy. Where there is an official registration procedure for disinfectant products only approved products should be used at the recommended dosage rate. Careful attention should be given to feed bins, watering devices and water lines to be sure that these are free of disease agents. Water lines should be flushed and then a disinfectant solution pumped into the lines. These lines should be closed and allowed to rest for at least 24 hours, and then thoroughly flushed to remove the disinfectant. An insect control programme should be implemented in conjunction with disinfection with a view to controlling both parasites such as red mite and nuisance insects such as darkling beetles.

Principle 5

Precautions must also be applied to major inputs, where practical. The major inputs are chicks, litter (where used), feed and water. For each of these a specification should be prepared, agreed with the supplier and regularly monitored.

Chicks

The chick specification should include details of the health status of parents with respect to vertically transmitted organisms (e.g. *Mycoplasma*, *Salmonella*, Chick Anaemia Virus, Avian Encephalomyelitis). Any requirement for negativity or positivity should be backed by a minimum frequency of sampling to confirm it is being achieved. The specification may also extend to general principles of hatchery and chick vehicle hygiene. Chick delivery is one of the most critical operations with respect to house biosecurity. It takes place at a time when external contamination of the site from a previous flock is likely to be at its highest.

Feed Hygiene

Feed is the largest input in terms of cost, and the largest in terms of bulk after drinking water. It comes into direct and intimate contact with the stock so it is important that it be both nutritionally and hygienically satisfactory. Feed mills should have a comprehensive and documented system for the production of safe animal foodstuffs, compliant with local legislation and Codes of Practice. This will usually involve an HACCP procedure with a view to avoiding unintended medicinal or feed-additive residues, and control of *Salmonella*. The plan should take into account the following:

Raw materials – Side-by-side with the traditional assessments of quality, feed materials should be sourced with a view to minimising contamination with poultry and zoonotic pathogens. The origin, transport, storage, processing and handling of feed material must be considered as part of this process. Store hygiene is particularly important since deficiencies in this area have been associated with contamination of raw materials with *Salmonella* (e.g. of rodent origin) and even of Paramyxovirus 1 (Newcastle Disease – of pigeon origin). Store management should be verified by annual inspection and/or membership of a recognised Assurance Scheme.

Some feed materials, such as soya bean meal, are processed prior to arrival at the feed mill. The majority of feed material will arrive in the mill unprocessed and direct from the farm or intermediate storage. Feed raw materials may be processed to improve the nutritional quality of the material or to reduce undesirable bacteria, such as *Salmonella*. The conditions of such treatment need to be carefully controlled (temperature, time, humidity) to consistently achieve the desired result. Excessive heat treatment may also reduce the nutritional quality of the material. Any material known to be contaminated with *Salmonella* should be put through a heat or chemical treatment to destroy the micro-organisms before the material is used for animal foodstuffs. Care must also be taken to avoid recontamination, for example, by contact with untreated material.

Finished feed – It is common practice for finished feed to undergo a final decontamination either through a high temperature process or through the use of chemicals such as organic acids. Once again, the principle of separation of treated from untreated feed materials must be used in order to reduce the risk of reinfection. Key aspects in this are limiting personnel access, filtering air to coolers and ensuring that finished products are kept well clear of incoming feed material and any air which might contain dust from these materials. Cooling systems require special attention because they encompass a range of physical environments with respect to temperature and humidity that can facilitate growth of moulds and bacteria. They also have a high demand for air. Just as for the raw materials it is important that the treatment regime does not damage the nutritional quality of the diet. Furthermore finished feed production, processing and handling systems must de designed and operated to avoid cross contamination of diets with feed additives and medicines.

Transport and delivery – Feed materials or compound foodstuffs may be readily contaminated if placed in contaminated vehicles. Vehicles used for carrying feed materials and finished foodstuffs must comply with appropriate legislation and Codes of Practice for Road Haulage. Particular attention must be paid to vehicle hygiene and cleanliness, correct loading, avoidance of contamination and cross-contamination, and delivery to correct farm facilities so that the foodstuff is received by the correct livestock. Scheduling of vehicles may be important where the mill is supplying farms of differing biosecurity status.

Vehicles may be contaminated from the general environment (e.g. road spray), farm environment (when loading or unloading), or from the transport of raw materials ('backloading'). These risks need to be evaluated for the specific feed production system, and they need to be managed with a balanced approach to vehicle dedication, maintenance and cleaning. Vehicle drivers must not enter poultry houses. Ideally, feed should be transferred to bins on site without vehicles having to go within the biosecure area.

Farm storage – Bins can harbour a range of bacteria, moulds and even coccidial oocysts. Bin hygiene may be substantially influenced by the design of the installation to reduce air and dust contamination within the house. Condensation (relating to temperature of delivered feed), and poor weatherproofing will also strongly influence conditions for microbial growth in the bins. It should be kept in mind that conditions within the bin may support the growth of organisms that have direct relevance to bird or human health (*E. coli* and *Salmonella* respectively). Also mould growth may result in production of mycotoxins that, even if they do not cause typical disease, can have a substantial effect on productivity. Smooth bin surfaces and access for inspection, and for cleaning if required, are especially critical at farm-depletion to ensure that bins and augers are not a source of contaminated material for the next flock. Natural day/night temperature cycles in a house between flocks can cause caked materials within the feeding system to flake off. Running augers prior to filling the system with feed will help monitor system cleanliness and help improve it. Some companies routinely treat feeder systems with a couple of kilograms of feed acidification product prior to filling. If wet cleaning is practised in the area of the feeder system it is important to ensure that drainage holes are located to allow complete drying of the system before any feed material is put through.

Treatments – A number of products, mainly based on mixtures of organic acids and their salts, but sometimes incorporating an aldehyde, are commonly used to reduce bacterial numbers and prevent their growth in feed. Complete decontamination of feed using these products alone is difficult to guarantee. However, many of these will continue to act for some time after their application and can, potentially, improve the general hygiene of the feeding system right the way through to the bird feeder. They should not, however be regarded as a substitute for the hygienic measures noted in the previous sections.

Drinking water hygiene

Drinking water is the largest single input into any poultry production system. It should be potable i.e. of a quality suitable for human consumption. However this alone is not enough. The importance of effective cleaning and sanitising of drinker systems at farm depletion has been emphasised above. Nipple drinkers are the preferred type. Particular drinker systems (e.g. 'bell type') are, by their nature, prone to bacterial contamination from the air. All drinker systems are prone to microbial growth when exposed to high environmental temperatures and low flow rates. These conditions apply especially during the first 1–3 weeks of life of young poultry.

Even if chlorinated public water supplies are used, the residual chlorine will be insufficient to control the risk of contamination. Dosing with approved water sanitisers during the first few weeks, during periods of disease challenge, or even throughout the life of some classes of poultry will help minimise this risk. Care must be taken with dosage to ensure effective doses without reducing water intake. If using chlorine-based treatments, residual levels of 1–5 ppm at drinker level have been recommended. Simple colorimetric test kits are available to estimate residual chlorine levels. This is necessary because the amount of chlorine required will be affected by the physical nature of pipework and the degree of accumulation of biofilm or other organic matter on surfaces.

Mild acidification of water lines may be used as an alternative to chlorination (it should certainly not be done at the same time). This is especially useful for cleaning water systems prior to the use of medication or vaccination. It is vital that no water treatments should interfere with vaccines applied in drinking water. Consult your poultry veterinary surgeon and/or vaccine manufacturer for specific advice in this area.

Litter

In deep litter systems, any litter introduced, whether at the beginning of the flock or in relittering, is a potential source of contamination with disease-producing organisms. In general terms wood shavings are less likely to present a bacteriological hazard than untreated straw. Contamination with spores of mould (usually *Aspergillus fumigatus*) can cause disease in young chickens, and in turkeys of any age. It occurs when litter materials have been high in moisture content and exposed to warm temperatures. Even if temperatures subsequently drop and the material dries out, large numbers of spores will persist. If litter is to undergo a treatment process then similar concerns about separating treated and untreated material apply as for feed and raw materials. Finally the process of storage, transport and delivery of litter into the poultry house should be reviewed with a view to avoiding recontamination. The weakest link in this chain is likely to be the actual delivery of litter into the poultry house. The external environment of the farm is likely to have some contamination from the previous flock when this is taking place.

Dead bird disposal

It may seem strange to include dead bird disposal in a section dealing with major inputs. Although they are actually outputs of the unit concerned they have the potential to serve as an input in another site if the disposal system does not work properly. This might be because wild animals gain access to disposal pits (where these are an approved method), or leakage from containers where a collection system is in place. Successful methods of dead bird disposal must prevent spread of pathogens to surviving birds, contamination of surface or ground water, and risk to human health. Several methods are acceptable in commercial systems. Strict biosecurity rules need to be applied to any system involving routine collections of dead birds from different sites. This is usually achieved by having a dead bird collection point at the edge of, or outside, the biosecure zone.

Disadvantages of Biosecurity Procedures

The standard recommendations for biosecurity as described here are broadly beneficial for safeguarding the health, productivity and welfare of commercial poultry. As is often the case with precautions, they can introduce some unintended consequences however.

Single-age effects

The cost of establishing biosecurity and satisfying other regulatory requirements are factors that tend to increase the number of birds per house and per site. The recommendation that they be of a single age means that when disease challenges do occur there is a large population of birds of similar susceptibility. This may exacerbate some disease conditions.

Time of infection effects

Certain infections cause minor disease when they occur in breeding birds in rear (e.g. Chick Anaemia Virus, Avian Encephalomyelitis). These infections can occur naturally but with improved biosecurity they will tend to occur later in life. If they occur during the laying phase they can cause chick quality problems and frank disease. Known infections that behave in this way are countered by vaccination in rear (see sections dealing with CAV and AE).

Group Tasks on Biosecurity

This section provides examples of tasks appropriate as group exercises during staff training on biosecurity. Where possible such tasks should be adapted to be as meaningful as possible for the trainees, perhaps by using examples of particular farms and particular hypothetical or real disease problems. After working through Task A you may wish to look at one suggested answer in Appendix 2.

Task A – Depletion/Turnaround

Having regard for biosecurity objectives take the case of a 15 000 bird broiler parent laying site, review the following list of activities and place them in a chronological order. What would you do first, second, third etc? The history for this particular flock was that a positive *Salmonella* Enteritidis isolate was obtained from the hatchery during the week of depleting the previous flock. Government authorities declined to take carcases to establish their true infection status. The list of activities is currently totally random. Be prepared to explain your reasoning. Feel free to add other activities to the list. Which ones have we missed out? It might be appropriate to include some activities more than once in your list.

Poultry Diseases Pocket Guide

Activity	Order	Why?
Order shavings		
Cut weeds		
Remove birds		
Clean and disinfect staff room		
Order feed		
Re-bait rodent boxes		
Set up equipment		
Disinfect house		
Staff holiday		
Close up cracks under doors		
Blood-sample birds for S.E.		
Wash house		
New or laundered overalls in use		
Sample litter for *Salmonella*		
Receive pullets		
Spray house with insecticide		
Fix broken wall boards		
Remove litter		
Install foot-baths		
Clean and disinfect bait boxes		
Remove debris around the houses		
Re-wire fans (Health & Safety)		
Inspect house		
Cement up cracks in dwarf walls		
Fumigate house		
Wash nests and slats		

Chapter 2 - Principles of Biosecurity

Task B – Routine Management

Taking the following broad headings of biosecurity hazards for a broiler parent flock in rear, identify the various components in each category, indicating specific on-farm action required for their control.

Main bio-hazard categories	Sub-divisions	On-farm action?
1. Previous flock	_____	_____
2. Chicks received	_____	_____
3. Other animals	_____	_____
	_____	_____
	_____	_____
4. House environment	_____	_____
5. Farm environment	_____	_____
6. Farm staff	_____	_____
7. Visitors	_____	_____
8. Equipment	_____	_____
9. Bedding	_____	_____
10. Feed	_____	_____
12. Litter	_____	_____
13. Water	_____	_____

THE IMMUNE SYSTEM: VACCINES AND HOW THEY WORK

CHAPTER 3

THE IMMUNE SYSTEM: VACCINES AND HOW THEY WORK

When a bird is vaccinated, or exposed to a viral or bacterial infection, a complex biological mechanism is set in motion that normally results in the elevation of the bird's specific defences against the disease in question. Sometimes this process also raises the bird's nonspecific defences against other infections. The immune response is generated by a complex system of specialised cells, the lymphocytes. All vertebrates have such a system but those of birds and mammals are the most complex. Standard serological tests measure only one component of the immune response, the antibodies circulating in the blood. Antibodies are proteins with one or more binding sites that attach to a specific site on a pathogen. The other main components of the immune system, which are not measured by standard serological tests, are antibodies produced and secreted locally (in tears, tracheal mucus, on the intestinal mucosa etc.), and the cellular immune response or delayed hypersensitivity.

Hieronymus Fabricius described the location and structure of a diverticulum of the avian cloaca in the late 16th century. It took almost another four centuries before the fundamental significance of the bursa for the development of immunity in birds was recognised. It was found that cells developing in the bursa and those developing in the thymus had different functions in the immune response. Both thymus and bursa have a role in producing or controlling the production of the antibodies that we measure in serological tests. The separation of the two central maturing organs that is present in the fowl has led to its use as a model for the investigation of many basic immunological phenomena. The antibodies which are secreted at mucosae are designated Immunoglobulin A (IgA), while Immunoglobulin M (IgM) and Immunoglobulin G (IgG) circulate in blood and lymph. The classes of antibody have varying chemical structure and numbers of attachment sites per molecule. Serological tests also vary in their ability to detect the different classes. Once the initial challenge has been dealt with, a group of cells (the so-called 'memory' cells) that have the required genetic make-up to produce antibody against the specific antigen, remain. Five days or so are generally required for the immune system to respond to the initial challenge but these cells allow a much more rapid and vigorous response to the secondary stimulus. This is known as an 'anamnestic' reaction.

Immunisation

Vaccination against primary viral pathogens helps reduce the need for all types of antimicrobial medication. Facilitation of the licensing of a broad range of cost-effective vaccines, which are safe and effective under field conditions, is the measure open to the regulatory authorities that is most likely to reduce the need for therapeutic antimicrobials and, hence, the risk of resistance development. The deliberate induction of immunity by vaccination is far preferable to natural induction after unpredictable exposure to field infection. Numerous infections, sometimes in combination, can kill or debilitate susceptible poultry causing pain and suffering in addition to losses in performance. Immunity is of two broad types: passive or active.

Passive immunity occurs as antibody in the yolk of developing embryos, derives from the maternal bloodstream and is present for 2–4 weeks in the blood of newly hatched chicks, until metabolised. Passive immunity is generally effective against viral diseases, but less so or ineffective against bacterial infections, e.g. mycoplasmas or salmonellae.

Active immunity occurs when an antigen is introduced to the bird and processed through the bird's immune system, resulting in various protective responses that will act to protect the bird if it is re-exposed to that antigen. Active immunity can be produced either by living or inactivated antigens, or a combination of the two. Live vaccines can be administered either to individual birds, such as by injection or eyedrop, or to large numbers of birds via the drinking water or by aerosol.

Inactivated vaccines must be given by injection. These usually incorporate potent adjuvants that enhance the local cellular reaction and, therefore, increase the immune response.

Immunity against some infections can be induced by injection of vaccine into the egg shortly before hatching, so that active resistance is developing before any exposure can take place.

Development of immunisation programmes

The development of an immunisation programme should be based on knowledge of the diseases to which birds are likely to be exposed and incorporated into the management system of the flock. For some infections it requires knowledge of the presence and level of passive immunity so that immunisation can be properly timed. Timing is also important so that vaccines do not detract from each other's responses or exacerbate their clinical effects.

Vaccines should not be administered when other stressors are acting on the flock. Immunisation cannot be a substitute for proper sanitation and biosecurity and programmes cannot totally protect birds which are stressed or in unhygienic conditions. Vaccines should be purchased and utilised only after full consultation with a poultry veterinary surgeon. Where monitoring tests are available, e.g. serology, these should be routinely used to ensure that vaccine responses have taken place. Vaccination programmes are a major component in veterinary health plans that are discussed in more detail on page 56. They should document the product to use, manufacturer, age administered, dose and route of administration.

POULTRY MEDICATION: A PRACTICAL GUIDE TO MEDICINES AND FEED ADDITIVES

CHAPTER 4

POULTRY MEDICATION – A PRACTICAL GUIDE TO MEDICINES AND FEED ADDITIVES

The treatment of disease in the modern poultry unit can be complex, with a range of medicines available for a variety of conditions. In most countries the majority of medications used in poultry are 'prescription-only medicines' (POMs). You are advised to consult your veterinarian when assessing or making use of information given in this chapter.

Decisions to medicate poultry must nearly always be based on the population, the flock, rather than the individual. Individual medication may be appropriate when small numbers of birds are affected, particularly if they are of high monetary value, such as breeding birds.

We need to carry out most medications of poultry on a flock basis for a number of reasons:

<u>Iceberg effect.</u> The poultry flock suffering a disease challenge is analogous to an iceberg, as shown in figure 5. Sick birds may be discernible with careful inspection however there is no direct correlation between discernible sickness at the individual bird level and benefit from medication. In many diseases the birds which are clinically ill respond poorly to medication because of the advanced state of their illness, while others which have not yet developed clinical signs respond to prompt effective medication.

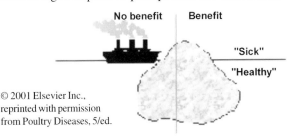

© 2001 Elsevier Inc., reprinted with permission from Poultry Diseases, 5/ed.

Figure 5

Challenge reduction. Because of the intimate contact between large numbers of individuals in a poultry flock, control of infection even in birds that would not become sick, may be of benefit by reducing the exposure of flock-mates to it. This is the basis on which many prophylactic or preventative treatments, for instance for *Mycoplasma* infection, work. It should be kept in mind that the number of bird-to-bird routes of infection (RI) rises dramatically as numbers increase (the formula is $RI = N \times (N-1)$ where N is the population size). For typical sized commercial poultry flocks the number of routes will be in the range of 10 million to 1 billion!

Practicalities. The practical difficulties of individual treatment of birds in a poultry flock are often not appreciated by people outside the industry. Assuming an average sized broiler flock of 20 000 birds and that one person catching could service two people evaluating/treating then about thirty people would be required to treat a flock once in one day. Frequently more than one house will be affected at the same time and treatment will be required at short notice making it virtually impossible to get the required number of people. In meat-type birds it is usually not possible to use long-acting medications because of the need to have short withdrawal periods before slaughter, so repeated treatments would be required. It is unlikely that any more accurate dosage could be achieved because it would be necessary to set a standard dosage irrespective of individual body weight. There are also serious welfare concerns with respect to ensuring that the handling and treatment do not result in stress and injury. Penning of birds to catch for treatment will cause both stress and facilitate the spread of infection.

Legal Requirements

In Europe these are based on EU directives. Medicines must be used safely and correctly in food-producing animals to ensure there are no residues. Most countries across the world have strict controls over both the methods of prescribing and the uses of medicines. In the EU for example medicines may be considered in five groups, which are adapted in the UK as follows:

Chapter 4 - Poultry Medication

GSL: **G**eneral **S**ales **L**ist which includes a variety of medicines that are available to the general public over the counter.

P: Medicines that are available over the counter only from a qualified **P**harmacist where his expert advice and guidance can be given or supplied by a veterinarian for animals under his care.

PML: **P**harmacy **M**erchant **L**ists: Certified merchants who are able to sell from a prescribed list of drugs direct to the farming community.

POM: **P**rescription **O**nly **M**edicines only available on the direction of a veterinarian.

Controlled Drugs: This category covers the addictive drugs such as morphine, heroin and pethidine. These are not used in poultry medicine.

Feed Additives or Zootechnical Additives. These are currently controlled by separate legislation in the EU although some (e.g. for the prevention of coccidiosis in chickens) have medicinal functions.

The medicines legislation is evolving in many countries towards increased control at the levels of manufacturer, distributor and user. The EU has recently published new feed additive regulations and is negotiating changes in the way the distribution of veterinary medicines is regulated.

The major categories concerning the poultry farmer are the PML and POMs.

When a medicine is supplied to the farm the following information should be available:

- A description of the medicine.
- The date of manufacture.
- The date of dispensing.
- The date of expiry.
- The client's name and address.
- The species to be treated.
- The specific birds to be treated – usually identified by house.

- The period of withdrawal required.
- The dose rate and instructions for use.
- Name and address of the supplier.
- Manufacturer's batch No.
- The name and address of the veterinarian prescribing.

How Medicines are Prescribed

Most medicines have two names, one which describes the chemical which is the active principle, often referred to as the **generic name**, and the second, the manufacturer's own **trade name**. For example, amoxycillin trihydrate is the generic name for a broad spectrum antibiotic. Vetremox is one of the trade names given to it, in this case by Alpharma Ltd. Other trade names for the same drug include Amoxinsol, Amoxypen, Clamoxyl etc. These vary from country to country.

Understanding Dosage Levels

All medicines have a recommended therapeutic range, and this is usually expressed in milligrams per kilogram (mg/kg) of live body weight. This range is used by the veterinarian so that he can decide whether a higher or lower dose level is required. In some circumstances the dosage may be specified as a concentration of the product in drinking water (e.g. grams/200 litres) or in feed (usually grams/tonne or ppm).

For information:

1000ng (nanograms)	= 1µg (microgram) also mcg
1000µg	= 1mg (milligram)
1000mg	= 1g (gram)
1000g	= 1kg (kilogram)
1000kg	= 1T (tonne)
mg / kg = g / tonne	= ppm (parts per million)
mg / kg x 0.0001	= %
ppm x 0.0001	= %
1000µl (microlitres)	= 1cc (cubic centimetre) or 1ml (millilitre)
1000ml	= 1 litre

Chapter 4 - Poultry Medication

Administering Medicines

There are three main ways of administering medicines to poultry.

1. **By injection** – The injection can be either intravenous (rarely in poultry), subcutaneous, intramuscular or by stabbing (e.g. of wing web or foot web).
2. **Topical** – The medicine is applied to the surface of the body.
3. **Oral** – The medicine is administered through the mouth. Most injectable antibiotics are also available for oral administration. They may be administered either by drinking water or in the feed.

Self Inoculation – What to do

If you inoculate yourself accidentally you should take the following actions.

1. Report immediately to your manager; ring your veterinary surgeon and/or your doctor.
2. Look at the label on the bottle. Does it give any emergency procedures?
3. Read the leaflet or data sheets that should be held on the farm.
4. If you are using an oil-based vaccine (see the bottle label) go to the casualty department of a hospital immediately with the bottle. Such vaccines can cause blood vessels to go into spasm. Vaccine manufacturers can provide an accidental vaccination medical information card to be presented to the attending doctor.

Administering Medicines Orally

Water Medication

Treating poultry via the drinking water involves a daily intake of the active principle based on mg/kg of live weight. In practice however, it is better to consider this by the amount of active drug required per day per tonne of live weight of poultry. Water medication can be provided in a short period of time and be given easily and quickly in the early stages of disease. It also allows large numbers of animals to be treated at low cost, however it can also be wasteful in that it is, of necessity given to healthy birds that do not need it.

Key Things to Know

- Sick birds often don't drink, and usually do not eat.
- The design of the pipe system may not be suitable.
- Water pipes tend to block up.
- Water tanks that are small require regular administration.

Applying antibiotic powder to water in header tanks

1. Calculate the total kg of liveweight to be medicated in tonnes.
2. Calculate the weight of powder required for the 24-hour period.
3. The water intake per tonne of liveweight per 24 hours is usually in the range of 100 to 300 litres. Young birds in brooding phase drink proportionally more, and heat stress can result in dramatic increases in water intake. Water intake on bell drinkers can be 5–10% higher than nipples with drip cups or trays.
4. Calculate the total water used in 24 hours (ideally based on recent water meter readings).
5. Divide the header tank capacity into the total water used which gives the times that the tank is emptied in 24 hours.
6. Divide the powder and add pro rata to the tank. Stir each time.

Example: Medication with Amoxycillin – 100% at 20mg/kg/day. A single 75gm pot is suffient to treat 3750kg of liveweight for a day (75000mg/20). If we have 5000 birds weighing 1.2kg that we wish to medicate for 3 days the total weight to be treated is 18 tonnes (1.2kg x 5000 birds x 3 days). We therefore require 4.8 pots (18000 kg/3750). It is normal to round up to 5 pots in this circumstance. The daily dose would be 1.66 pots (5pots/3days). Place 62.5g or half the powder in the tank early in the morning and the other half in the afternoon. For certain products your veterinarian may recommend 'pulse dosing' in which the medicine is applied over a shorter period each day, either on a daily basis or perhaps morning and evening. Both continuous and pulse dosing have their advantages and the choice depends on the disease being treated and the product being used.

Dissolving Powders – Useful Tips

- ❖ Place the required amount of powder in a dry bucket.
- ❖ Fill another bucket with water.
- ❖ If the powder is poorly soluble it is best if the water is lukewarm.
- ❖ Tip the water in one go into the dry bucket, immediately stirring the powder into solution.
- ❖ When choosing automated dosing equipment ensure that it can handle proportions up to 5%.

Administrating Medicines In Feed

The inclusion of antibiotics in feed is a common method of controlling and preventing diseases. In-feed medicines are prescribed by grams (gm) of active or generic substance per tonne of feed. However, the manufacturer's product is normally available as a supplement – a mixture of the generic substance and, usually, a cereal base. For technical and analytical reasons do not expect a 100% recovery rate if the feed is analysed. It is usually 80–90% but can be as little as 40–60%.

Key Things to Know

- The bin containing the medicated feed should be marked with the fill and empty dates to allow withdrawal times to be calculated.
- If medicated feed is placed in a bin containing non-medicated feed, the time and quantity of the medicated feed reaching the birds will be unknown.
- There can be a delay in manufacturing and delivering medicated feed.

Poultry Diseases Pocket Guide

- Sick birds often do not eat or have reduced feed intake and therefore may not receive sufficient antibiotic.
- In-feed drugs require a product licence for use in food-producing animals and therefore the availability of drugs is limited.
- A Medicated Feedingstuffs Prescription or equivalent document is required from your veterinarian in many countries. You should find out whether you need to request this or will the feed mill do it for you.

Further information relevant to in-feed medication is presented in following sections dealing with residue avoidance and in-feed medication and feed additives.

Treating Individual Birds

This is occasionally appropriate with high-value breeding birds. When you are considering treating birds orally (or by injection) ask yourself the following questions:

- Should I consult my veterinarian?
- Have I identified every individual affected bird?
- Is this condition one that has been reliably diagnosed before or is it a new one?
- Is it necessary to treat it?
- Do I have drugs to treat this condition or are they readily available?
- Are there any welfare or nursing implications?
- Should the affected bird(s) be moved to a hospital pen?
- Is injection the best or should I use a different method of administration?
- What dose should be given? Have I the right information on this?
- How often should the drug to be given and for how long?
- Are any adverse effects likely?

Then you should:

- Record when the treatment started and its progression.

Chapter 4 - Poultry Medication

- Assess the response on a day-by-day basis.
- If there is no response within 24 hours consult your veterinarian.

Other Types of Medicine for use on the Farm

Vaccines

Examples of available poultry vaccines will be detailed in a companion book. Their availability varies from country to country.

Electrolytes

When a bird looses fluids due, for example, to diarrhoea it looses elements (electrolytes) such as sodium and potassium, and hydrogen ions. These must be replaced. As little as 7% loss of body fluids causes marked clinical signs, 15% loss of body fluids will cause death.

Antimicrobial Medication

Most medicines used in commercial poultry are antimicrobials or antibiotics. Antimicrobials are chemically synthesised and antibiotics are produced by fermentation but both are used to treat bacterial infections. If a significant number of animals in a group show signs of disease, both sick and healthy animals may need to be treated with therapeutic levels of a product for the recommended period. This is intended to cure the clinically affected animals, reduce the spread of disease and prevent clinical signs appearing in the remainder. Table 1 below lists diseases of poultry most commonly requiring medication. Medication should always be used at the dosage and period of treatment recommended by your veterinarian.

Flock medication may be applied in the drinking water or in the feed. Some products may be applied as a 'pulse dose' by giving, say, half of the total amount required for the day for a short period in the morning, and the remainder in the afternoon.

Poultry Diseases Pocket Guide

Table 1 Examples of poultry diseases requiring medication with compounds with antimicrobial activity

Chicks: - First week septicaemia. - *Mycoplasma* infection. Broiler chickens: - Septicaemia due to *E. coli*. - Osteomyelitis/femoral head necrosis. - Necrotic enteritis.	Broiler chicken breeders: - *Staphylococcus aureus*, joint infections. - *Pasteurella*. - *Mycoplasma*. Turkeys: - *E. coli* septicaemia following TRT. - *Pasteurella*.

Drinking Water Medication

Advantages:

- Can start quickly regardless of feed stocks
- Sick birds continue to drink even when they are not eating
- Can apply as pulse or continuous medication easily
- It is easy to determine the end of medication
- Feed residues in bin are not a problem

Disadvantages:

- More expensive per bird treated
- Some products are poorly soluble
- Some products may cause line blockage or valve stickiness

In-Feed Medication

Advantages:

- Lower cost per bird treated
- Convenient for routine prophylactic treatment

Disadvantages

- Feed bins may be full when treatment required
- Care required in mill and with bins to avoid cross-contamination of products
- Unless separate bins are available for medicated feed it can be difficult to say exactly when treatment finished
- The mill requires a Medicated Feedingstuff Prescription from your veterinarian before delivering the feed

Chapter 4 - Poultry Medication

Treatment of clinical disease

Coli-septicaemia, either acute or subacute, is a common sequel to a number of viral diseases of poultry, especially those caused by viruses of the respiratory system. This is one of the more common reasons for medication. Samples of ill or dead birds should be routinely subjected to post-mortem examination to confirm the diagnosis and, commonly, to isolate and sensitivity-test the pathogen. It is a good idea to have the diagnosis done at an early stage in the disease process so that this result can help guide the treatment.

Prevention of clinical and subclinical disease

Some bacterial infections are best dealt with by treatment before the clinical signs. Many such infections have been eliminated from the majority of the poultry population by means of eradication programmes (e.g. *Mycoplasma gallisepticum, M. meleagridis, M. synoviae*). However there is a reservoir of these infections in 'backyard' flocks, game birds and, possibly, wild birds. 'Breaks' can be expected, and, in this circumstance it will be common to treat the affected parent flock and the progeny with an appropriate product.

Responsible use of medicines

In order that these medicines continue to work effectively in your poultry and that their use causes no adverse effects for human health it is important that all medicines are used responsibly. Guidelines to help us achieve this aim have been published in many countries. The Responsible Use of Medicines in Agriculture (RUMA) Alliance has published recommendations and advice that you may find helpful (see www.ruma.org.uk). On-farm quality assurance schemes may require that you have a copy of these and similar guidelines. Your veterinarian is also expected to operate within guidelines. Those published by the British Veterinary Poultry Association (www.bvpa.org.uk) may be summarized as follows:

Poultry Diseases Pocket Guide

Therapeutic antimicrobial products should:

- not be used as an alternative to good management, vaccination, or site hygiene.
- only be prescribed for animals under care of attending veterinarian.
- not be used long term in absence of disease.
- be used in accordance with sensitivity of causal organisms.
- be used in prophylactic medication only in accordance with agreed policy of practice.
- be used in such a way as not to adversely affect the documented preventative medicines programmes.
- be used in a dosage regime to maximise efficacy while minimising resistance.

Usage monitoring schemes should not hinder prevention of suffering and should take into account potency of compounds used. The simplest approach is to record the number of kgs of animal treated/day as a proportion of the total kgs of animal at risk.

Recording of Medicines Use

In most countries it is a legal requirement to maintain careful records of all medicines used in food-producing animals. When the medicine is supplied you should be given, on a label, prescription or data sheet, information of the type shown below. The format may be different. Some major purchasers of poultry products and quality assurance schemes require that a prescription be issued for any use of antimicrobials.

Regardless of the documentation obtained from your veterinarian it is important that you keep a good record of actual on-farm usage in a consistent format. This will help you ensure that all withdrawal periods are met and you can then document your use of medicines.

Chapter 4 - Poultry Medication

SPECIALIST VETERINARY SERVICES   Telephone 01XXX XXXXX
                                  Fax:01XXX xxxxx
Veterinary Laboratory, Back Lane, Hemingborough, Anyshire

PRESCRIPTION   Ref: RH.96.0108

To: Mr J. Smith   cc. Woodside Farm
    Big Chickens Ltd
    The Hatchery
    Upper Broadlands
    Littleton
    Anyshire AN23 7XY

Site: Woodside   Broiler Chicken   House(s): 3
Supplied: Vetremox - 23 x 75 gm. pot(s) from store:D
For Animal Treatment Only
This prescription documents the supply of medicine for birds under our care
Vetremox   Product License: 11003/4000   Legal Category POM
Dosage Rate: 20mg/kg amoxycillin, use one 75 gm. pot per 3750 kg per day.
Route and Course of treatment: Drinking water for 3 days. The requirement cf 23x75 gm. pot(s) of Vetremox was calculated on the basis of 3
days treatment of 23 000 birds weighing 1.2 kg each. Use approximately 7.7 x 75 gm. pot(s) per day.

IMPORTANT: Human Food Residue Avoidance Information.
The treated birds may only be slaughtered for human consumption after 24 hours have elapsed after the end of the period of treatment.

Figure 6 – An example of a medicine prescription for poultry.

A FORMAT FOR RECORDING POULTRY TREATMENTS

Date Started	House	Condition/ disease	Medicine	Units /day	Unit size	Withdrawal (Days)	Date withdrawal completed	Administered by
1-8-03	3	Necr.Enter.	Amoxypen	1	250gm	1	5-8-2003	JD

Poultry Diseases Pocket Guide

It goes without saying that all medicines used in food animals should be purchased in accordance with the local legislation. This will normally require that the product is licensed by a national regulatory agency and distributed through an approved distribution chain. Purchase of irregular product may risk prosecution, poor product quality and lack of efficacy.

Controlling and Storing Medicines

To achieve the maximum response to medicines and prevent any abuses, discipline should be maintained in their control, administration and storage. Consider all drugs to be dangerous. Many become potentially toxic if the recommended levels of treatment are exceeded or if they are given in the wrong way.

Light and heat destroy drugs and freezing also has an adverse effect, particularly on vaccines.

A check list

- ✓ Provide a locked room or cupboard for all your drugs. Store all antibiotics in closed containers in the dark at 10–22°C.
- ✓ Provide a refrigerator for vaccines and other drugs as required. Use a maximum minimum thermometer and record temperatures daily. Temperature should be in the range of 2–6°C. Most vaccines are damaged by holding at too high a temperature, or by freezing so it is wise to allow sufficient space for air circulation in the fridge, and store vaccines on a shelf separated from the cooling plate and freezer box (if there is one in the fridge).
- ✓ Allow only certain designated people to have direct access to the main drug store.
- ✓ Document all drugs in and out of the drug store.
- ✓ Insist on empty bottles being returned before a second bottle is taken out. This prevents black market trade.
- ✓ Agree with your veterinarian the minimum amounts that are required for a given period of time and follow his advice on usage.

Chapter 4 - Poultry Medication

- ✓ Make sure that all bottles are labelled for the correct use, that withdrawal periods are displayed and personnel are aware of them.
- ✓ Follow the instructions precisely.
- ✓ Keep a daily record of all medicines used on the farm.
- ✓ Make sure you have safety data sheets to hand in case of accidents.
- ✓ Ask your veterinarian to check your storage and usage of medicines regularly to ensure that the recommendations are being carried out.
- ✓ Always be aware of the withdrawal period before slaughter.
- ✓ Check regularly that medicines are in date.
- ✓ Make sure that all drugs, syringes and needles are kept well away from children and people not on the staff.
- ✓ Dispose of empty bottles, needles and syringes safely.

Drugs that should be stored in the refrigerator
2–8°C (36–46°F)

- All vaccines including part opened bottles.
- Any bottles that have been opened and are in use.
- Any other medicines where the label indicates this temperature requirement.

Drugs that should be stored in a dark, cool place
18–22°C (64–71°F)

- Vitamins and minerals.
- Antibiotics.
- In-feed and water soluble preparations.
- Disinfectants.

Disposing of Medicines

This must be carried out with care to prevent environmental contamination and accidental human or animal contamination.

- Empty bottles should be placed into a plastic bag and disposed of in line with local authority guidelines or rules.

- Needles <u>must always</u> be removed from the syringe, and the syringes placed in polythene bags, marked 'Syringes only' and incinerated.
- Bottles that have contained a live vaccine should be disinfected prior to disposal. Bleach is suitable for this.
- Needles and needle holders should be placed into a Sharps box, to be taken away for incineration.

Residue Avoidance and 'Annexe 4'

In addition to the foregoing advice with respect to withdrawal periods, it is worth noting the situation in Europe with respect to certain medicines. All medicines used in food animals in Europe are now required to have been assessed and given a Maximum Residue Limit (MRL). Some groups of compounds and individual compounds have been withdrawn because it was decided that it was not possible to set an MRL, and these are placed in 'Annexe 4' – a list of prohibited substances. Until recently this legislation applied only to medicines, not feed additives, but it is being extended to include these. New testing methodology has greatly improved the sensitivity of the detection of these compounds and their metabolites. In order to avoid serious trade restrictions it is vital that no food animals producing goods aimed at the European market are either treated with, or accidentally exposed to, Annexe 4 products. This currently includes, among others, all nitrofurans (furazolidone, nitrofurazone, furaltadone, nifursol), chloramphenicol, and the imido-thiazole group (dimetridazole, ronidazole etc).

Feed Medication and Feed Additives

All feed delivered to poultry sites should be accompanied by documentation that lists any medications and additives that it contains.

Medication may be prescription-only or non-prescription. If prescription-only, there is a requirement in many countries that a formal medicated feedingstuffs prescription is signed by a veterinarian. This imposes a legal duty on the veterinarian to

Chapter 4 - Poultry Medication

establish that the medication is necessary, that the dosage and period of medication is appropriate, and that instructions are given to avoid residues in the products produced. Such medication may be for the control of bacterial disease either therapeutically or preventatively (prophylactically). Medicines used in this way include antimicrobials and antibiotics and, for some species and in some countries, wormers (anthelmintics) and antiprotozoal products.

In some countries certain wormer and antiprotozoal medicines used in the feed do not require a prescription. In most countries anticoccidials and other antiprotozoal products, as well as digestive enhancing antimicrobials (growth promoters), are regulated simply as non-prescription medicines. In the European Union these products are covered by separate legislation, Directive 70-524, which regulates a broad range of feed additives throughout Europe. Re-registration of such products is currently ongoing and it is expected that only specific brands of such products will remain approved after this process is completed.

The debate continues on the use of digestive enhancers. The EU has announced that it wishes to suspend the two remaining approved products for broilers (flavomycin and avilamycin) with effect from 2005–2006. They have not indicated whether it is intended to apply the same ruling to products from other states or how they could monitor this. A significant benefit for animal welfare results from the improvement in the utilisation of nutrients and the reduction in the volume or moisture of undigested material deposited in the animal's environment. There are also beneficial effects for the overall environment with reduced feed such as fewer lorry journeys, lower water use and reduced arable land required to be planted in cereals. In addition to their direct economic effects, digestive enhancers also have benefits in the control of subclinical and clinical disease. This group of products can play a role in controlling the adverse effects of non-specific enteritis (of nutritional or viral origin) and in reducing the risk of necrotic enteritis and cholangiohepatitis.

Choice of feed additive programmes goes beyond the scope of this handbook. Most farmers will not be directly involved in these decisions. However they do have an important role to play in avoiding food residues. Some of the feed additives have withdrawal periods to be complied with. Careful bin management is required to achieve this. Anticoccidials containing nicarbazin require special care. The UK Veterinary Medicines Directorate has issued advice aimed at reducing the risk of nicarbazin residues (summarised in the box below). These principles are also valid for all feed containing medication and feed additives that have a required withdrawal period. Feed additive programmes, including withdrawal periods, should be incorporated in the 'Veterinary Health Plan' which you should agree with your veterinarian. This plan should also include any routine prophylactic medication in use.

Avoidance of Nicarbazin Residues – Advice from the VMD

- ❖ Deliveries of feed containing nicarbazin and feed not containing nicarbazin should be made in accordance with the United Kingdom Agricultural Supply Trade Association's Code of Practice for the Manufacture of Safe Compound Animal Feedingstuffs.

- ❖ Ensure that bulk bins holding feed containing nicarbazin are completely emptied out before refilling with feed not containing nicarbazin. Remember the movement of feed out of a bin occurs directly above the discharge opening. The remaining feed then cascades down the slope of the crater that is formed. Failure to completely empty bins before refilling may result in residual feed, which may contain nicarbazin, being left in the bin.
  *If possible there should be two feed bins. However, to prevent cross-contamination in units without two bins, clean out holding bins and delivery wagons before putting feed not containing nicarbazin into them.
  *Where two feed bins are used it is still important to remove all feed containing nicarbazin from a bin before feed not containing nicarbazin is added to the bin.

- ❖ Keep feed containing nicarbazin separate from feed not containing nicarbazin and clearly identified. If feed is removed from a bin, then full traceability must be maintained.

Chapter 4 - Poultry Medication

IF THERE IS ANY POSSIBILITY OF FEED CONTAINING NICARBAZIN, UNDER NO CIRCUMSTANCES SHOULD THIS BE FED TO BIRDS WITHIN 9 DAYS OF SLAUGHTER (OR 5 DAYS IF 50% COMBINATION WITH NARASIN).

- ❖ Always follow the instructions and observe the withdrawal period specified on the feed label or that given to you by your veterinary surgeon by removing feed containing nicarbazin for the appropriate number of days before slaughter.

- ❖ It is the responsibility of the farm manager to record all deliveries of feed and movement of feed between houses and maintain full traceability. They must be fully aware of the appropriate withdrawal period for each of their feed deliveries. Farm managers must ensure that the last possible exposure to diets containing nicarbazin at 100 ppm and 125 ppm is at least 9 days before slaughter (or 5 days if 50% combination with narasin is being used).

- ❖ Farm managers should be trained on all aspects of the use of feed containing nicarbazin, with an emphasis on the correct procedures for withdrawals prior to slaughter. Other relevant staff should have appropriate training to ensure they can assist the farm manager in their responsibilities.

- ❖ Any feed (starter crumbs etc) containing nicarbazin that is spilled into the litter must be cleared up immediately. This is necessary because chicks might not eat it straightaway, while they have other feed available, but may eat it when feed is withdrawn immediately prior to catching.

- ❖ Where growers are using a grower feed containing nicarbazin, it is important to remember that when they change to a finisher feed with a withdrawal period of 3 to 5 days, then the withdrawal period from the grower feed still applies.

- ❖ Do not remove birds for early slaughter when they are consuming feed containing nicarbazin. No birds must be allowed to go to slaughter until all withdrawal requirements have been met.

❖ We recommend that a 250 gm sample of withdrawal diet should be obtained from the feeding system and kept in a rodent-proof box for at least two crops.

If in doubt consult with your Feed Supplier and Veterinary Surgeon.

Veterinary Health Plans and Planning

A detailed health plan can be a very powerful tool to provide a framework for all health-related decisions and records. The health plan should be a collaborative exercise between the company or farmer and their veterinarian and it should summarise the key activities relating to health promotion. At Poultry Health Services we use a detailed system of health planning for flocks of parent chickens and turkeys because of the very many activities involved. For commercial broilers and turkeys we use either the companies own system or a HTML-based document similar to that used for transferring information on the Internet. As with any commercial activities we should be clear as to the objective of this:

1. To avoid misunderstanding or miscommunication as to what activities should occur and when
2. To facilitate recording of events as they occur
3. To facilitate the documentation of these activities to other interested parties such as customers and assurance scheme auditors.

There are two especially good reasons for you to put effort into health planning:

A. The various components of the plan can be additive in effect, synergistic (in which their sum is greater than the effects of each on its own), or antagonistic (one working against the other). For optimal economic results we aim for additive or, preferably, synergistic effects. In order to achieve this the plan must include the right components, done in the correct order and at an appropriate interval.

Chapter 4 - Poultry Medication

B. Because the interval between a change in the way we do things and its effects may be long, and because managers are often dealing with multiple flocks of different ages, it is difficult to remember exactly when changes were made. A series of accurate veterinary health plans can help document such changes and facilitate the accurate analysis of productivity and health information to generate improved knowledge of the problems, and eventually, help provide answers.

The degree of detail to include in the health plan is a fundamental decision and will vary according to the type of production, and its needs. Health plans that are out-of-date or inaccurate do not serve a useful function. We therefore advise that the degree of detail included is sufficient to achieve the objectives and does not involve any more administrative work than is necessary.

Here is a checklist to use in deciding what to include in your health plan:

1. By company, production area or by farm
2. Include a specific number or code to identify the plan
3. Identify date effective and identity of plan replaced
4. Name of Veterinary Surgeon
5. Name of farm or company manager responsible
6. Type of stock
7. Length of down-time between flocks
8. Cleaning and disinfection programme
9. Cleaning and disinfection contractor
10. Sanitation monitoring system
11. Hatchery of origin
12. Hatchery vaccines and interventions
13. On-farm interventions – operator, age, need
14. Vaccination programme on farm
    Vaccine name
    Vaccine dose
    Route of application
15. Competitive exclusion – hatchery or on farm

16. In-feed medications and additives
    Coccidiostats (or freedom from coccidiostats)
    Acidifiers
    Enzymes
    Added whole grain
17. Feeding programme – codes and ages
18. Health monitoring activities
    Criteria for post-mortem examinations
    Routine tests for zoonotic agents
    Routine tests for poultry disease

While this may seem a long list it is by no means exhaustive. One approach to simplifying the system is to move all items that are essentially fixed into a Health Policy document, leaving only those items that are adjusted periodically in the health plan itself.

One way of visualising and explaining the concept of the veterinary health plan is to liken it to a wall to protect the stock as shown in figure 7 below. The foundations are in the planning and implementation of new systems or refurbishments of existing ones. Taking into account health at this stage is likely to be the most cost-effective approach in the long run. The 'bricks' in the wall will vary according to the economic objectives, type of system and local disease challenge. 'Off-the-peg' plans are a useful starting point but need to be adapted to local circumstances and to changing patterns of disease.

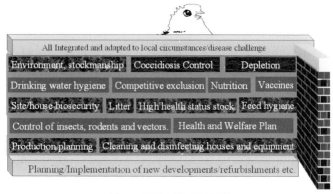

Figure 7 The Health Wall

Chapter 4 - Poultry Medication

Veterinary health planning is one of the major topics covered in 'Managing Poultry Health: The prevention and treatment of disease' which is currently in preparation.

DISEASES AND SYNDROMES: INTRODUCTION

CHAPTER 5

DISEASES AND SYNDROMES: INTRODUCTION

In the following pages around 140 diseases and syndromes of poultry are summarised. For each condition a brief introduction is given describing its cause, general features, morbidity and mortality. Morbidity is the proportion or percentage of the birds which become ill in an outbreak, whereas mortality is the proportion or percentage of sick birds that die as a result of the disease. The 'signs' are then listed. A clinical sign is any abnormality that can be appreciated from simple external examination of the live bird. Post-mortem lesions are then listed. These are gross lesions (visible to the naked eye) that are detectable by dissection of the carcase at post-mortem examination (by necropsy). In carrying out a post-mortem examination it is advised to proceed in a systematic standardised manner to examine the main organs. More detailed examination (including microscopic examination of fresh smears for example) may be advisable depending on the nature of the condition, the clinical signs and the tentative diagnosis based on the initial examination.

Key important terms are detailed in the glossary at the end of the book however the most important tip in understanding the terminology is that '-itis' at the end of a noun means an inflammatory condition of the named organ, and most of these are of obvious meaning (if you know the technical name of the affected tissue or organ):

Airsacculitis – inflammation of the air sacs

Arthritis – inflammation of the joint

Conjunctivitis – inflammation of the conjunctiva – the membrane covering the surface of the eye

Enteritis – inflammation of the intestine

Endocarditis – inflammation of the endocardium – the inner lining of the heart

Poultry Diseases Pocket Guide

Gastritis – inflammation of the stomach

Hepatitis – inflammation of the liver

Nephritis – inflammation of the kidney

Pericarditis – inflammation of the pericardium, the membrane around the heart

Some are not quite so obvious in meaning:

Encephalitis – inflammation of the brain

Keratitis – inflammation of the cornea – the normally transparent membrane at the front of the eye

Panophthalmitis – inflammation of the whole of the eye

Perihepatitis – inflammation of the capsule of the liver

Rhinitis – inflammation of the nasal passages

Salpingitis – inflammation of the oviduct

Typhlitis – inflammation of the caecum

See Appendix 1 for further definitions.

The following summaries of poultry diseases are divided into three sections:

Chapter 6 Diseases affecting chickens predominantly or multiple poultry species

Chapter 7 Diseases affecting turkeys predominantly

Chapter 8 Diseases affecting waterfowl (ducks and geese) predominantly.

The following table shows an index of the conditions in these three chapters along with the species in which they have been reported either as a disease or an infection (X), or in which they are only rarely reported (R). Even if a condition has not yet been reported in a species this does not mean that it cannot occur in that species!

Chapter 5 - Diseases and Syndromes: Introduction

Occurrence of disease by poultry species - Index

Chapter 6 - Chickens and Multiple Species

	Chicken	Turkey	Duck	Geese	Pheasant	Quail	Guinea Fowl	Pigeons	Page
Amyloidosis	X		X						73
Ascites, 'water-belly', Broiler Ascites	X								74
Aspergillosis	X	X	X	X	X	X			75
Avian Encephalomyelitis Egg Drop	X	X			X	X			76
Avian Encephalomyelitis, Tremors	X	X			X	X			77
Avian Leukosis, Lymphoid Leukosis	X								79
Avian Leukosis (Sero-type J)	X								80
Avian Rhinotracheitis	X	X			?		X		82
Beak Necrosis	X	X							84
Bedbug infestation	X	X							84
Big Liver and Spleen disease	X								85
Biotin deficiency and fatty liver and kidney	X								86
Biting lice	X	X	X	X	X	X	X		87
Blackfly infestation	X	X	X	X	X	X	X		88
Botulism	X	X	X	X					89
Breast Blister	X	X							90
Caecal Worm	X	X	X		X	X			91
Calcium Tetany	X								92
Campylobacter infection	X	X	X	X	X	X	X		92
Candidiasis, Moniliasis, Thrush	X	X							94
Cannibalism, feather pecking	X	X			X	X			95
Capillariasis or Hairworm Infection	X	X	X	X	X	X		X	96
Cellulitis	X								97
Chicken Anaemia	X								98
Chondrodystrophy, Perosis	X	X	X						99
Coccidiosis	X	X	X	X	X	X	X		
Coccidiosis, Upper intestinal, *E. acervulina*	X								100
Coccidiosis, Mid-intestinal, *E. maxima*	X								102
Coccidiosis, Mid-intestinal, *E. necatrix*	X								103
Coccidiosis, *E.mitis*	X								104
Coccidiosis, *E.praecox*	X								105
Coccidiosis, Caecal, *E. tenella*	X								106
Coccidiosis, Ileorectal, *E. brunetti*	X								108
Colibacillosis, Colisepticemia	X	X	X	X	X	X	X	X	109
Contact Dermatitis, Hock Burn, Pododermatitis	X	X							111

Poultry Diseases Pocket Guide

Disease	Chicken	Turkey	Duck	Geese	Pheasant	Quail	Guinea Fowl	Pigeons	Page
Cropworms	X	X	X	X	X	X			112
Cryptosporidiosis	X	X	X		X				113
Dactylariosis	X								114
Degenerative joint disease	X	X							115
Depluming and scaly leg mites	X	X							116
Dysbacteriosis - non-specific bacterial enteritis	X	X							116
Egg drop syndrome 76	X								118
Endocarditis	X								119
Epiphysiolysis	X								120
Equine Encephalomyelitis (EEE, WEE, VEE)	X								121
Erysipelas	R	X	R	R	R	R	R		121
Fatty liver haemorrhagic syndrome	X								123
Favus	X	X							123
Femoral Head Necrosis, Osteomyelitis/chondritis	X	X							124
Fowl Cholera, Pasteurellosis	X	X	X	X					125
Fowl Plague, Avian Influenza-Highly Pathogenic	X	X	X	X	X	X	X	X	127
Fowl Pox, Pox, Avian Pox	X	X						X	130
Gangrenous Dermatitis, Necrotic dermatitis	X								132
Gape	X	X			X	X			133
Gizzard worms - Chickens	X								134
Heat Stress	X	X	X						134
Hemorrhagic disease, Aplastic anaemia	X								135
Hydropericardium-Hepatitis Syndrome, Angara disease	X								136
Impaction and foreign bodies of the gizzard	X	X							137
Inclusion body Hepatitis	X								138
Infectious bronchitis, IB	X				?				140
Infectious bronchitis, IB - 793b variant	X								142
Infectious bronchitis, IB egg-layers	X								143
Infectious Bursal disease, IBD, Gumboro	X								144
Infectious Coryza	X				R		R		147
Infectious Laryngotracheitis, ILT	X	R			X				148
Intussusception	X								149
Malabsorption Syndrome, Runting/stunting	X	X							150
Marek's disease	X	R							151
Mycoplasma gallisepticum infection, M.g.	X	X	X	X	X	X	X		153

Chapter 5 - Diseases and Syndromes: Introduction

	Chicken	Turkey	Duck	Geese	Pheasant	Quail	Guinea Fowl	Pigeons	Page
Mycoplasma synoviae infection, M.s. Inf. Synovitis	X	X							155
Mycotoxicosis	X	X	X	X	X	X			157
Necrotic Enteritis	X	X	X						159
Non-starter and 'Starve-out's	X	X							161
Oregon Disease, Deep Pectoral Myopathy	X	X							162
Ornithobacterium infection	X	X							163
Osteoporosis, cage fatigue	X	X	X						166
PMV-1 or Newcastle disease	X	X	X	X	X	X		X	167
Paramyxovirus PMV-2, Yucaipa Disease	X	X							170
Pendulous Crop	X	X							170
Proventricular worms	X	X	X	X	X	X			171
Pullet disease, Bluecomb, Avian Monocytosis	X								172
Red mite and Northern Fowl Mite	X	X							173
Respiratory Adenoviral Infection, Mild respiratory disease	X								174
Respiratory Disease Complex	X	X							174
Reticuloendotheliosis, Lymphoid tumour disease	X	X	X	X		X			176
Rickets (hypocalcaemic)	X	X	X						177
Rickets (hypophosphataemic)	X	X	X						178
Rotavirus infection	X	X			X		X	X	179
Roundworm, large - Ascaridia	X	X	X	X	X	X	X	X	180
Ruptured gastrocnemius tendon	X								181
Salmonella Gallinarum, Fowl Typhoid	X	X	X	X	X	X	X		182
Salmonella Pullorum, Pullorum disease	X	X	X	X	X	X	X		184
Salmonellosis, Paratyphoid infections	X	X	X	X	X	X	X	X	185
Salmonellosis, *S.*Enteritidis and *S.*Typhimurium	X	X	X	X	X	X	X	X	187
Salpingitis	X	X	X						190
Spiking Mortality of Chickens	X								191
Spirochaetosis	X	X	X	X	X				192
Spondylolisthesis, Kinky-back	X								193
Spraddle legs, splay legs	X	X	X	X	X	X	X		194
Staphylococcosis, bumble foot	X	X							195
Sudden death syndrome, flipover	X								196
Tapeworms, Cestodes	X	X	X	X	X	X	X		197
Tibial dyschondroplasia, TD	X	X	X						198
Ticks	X	X	X						198
Trichomoniasis		X			X	X			199

Poultry Diseases Pocket Guide

	Chicken	Turkey	Duck	Geese	Pheasant	Quail	Guinea Fowl	Pigeons	Page
Tuberculosis	X	X	X	X	X	X	X		200
Twisted leg	X	X							201
Ulcerative Enteritis, Quail disease	X	X			X	X			202
Vibrionic hepatitis, Avian Infectious Hepatitis	X								203
Viral Arthritis	X								204
Visceral Gout, Nephrosis, Baby Chick Nephropathy	X	X	X	X	X	X	X		205
Vitamin A Deficiency, nutritional roup	X								207
Vitamin B Deficiencies	X								208
Vitamin E deficiency, Encephalomalacea	X	X	X						210
Yolk sac infection, omphalitis	X	X	X						211

Chapter 7 - Predominantly Turkeys

	Chicken	Turkey	Duck	Geese	Pheasant	Quail	Guinea Fowl	Pigeons	Page
Arizona infection, Arizonosis, Paracolon infection		X							215
Chlamydiosis, Psittacosis, Ornithosis	R	X	X						216
Coccidiosis of Turkeys		X							217
Dissecting aneurysm, aortic rupture		X							219
Haemorrhagic enteritis		X							220
Hexamitiasis		X	X		X			X	221
Histomonosis, Blackhead	X	X				X			222
Leucocytozoonosis	R	X	X				X		224
Lymphoproliferative disease (LPD)		X							225
Mycoplasma gallisepticum infection, Infectious sinusitis	X								226
Mycoplasma iowae infection, M.i.		X							227
Mycoplasma meleagridis infection, M.m.		X							228
Osteomyelitis complex		X							230
Paramyxovirus PMV-3	X	X							230
Paramyxovirus PMV-6		X							231
PEMS/Spiking Mortality of Turkeys		X							232
Shaky Leg Syndrome		X							233
Transmissible enteritis, bluecomb		X							233
Turkey coryza		X							234
Turkey Rhinotracheitis (in rear)	(X)	X							235
Turkey Rhinotracheitis (Adult)	(X)	X							236
Turkey Viral Hepatitis		X							237

Chapter 5 - Diseases and Syndromes: Introduction

	Chicken	Turkey	Duck	Geese	Pheasant	Quail	Guinea Fowl	Pigeons	Page
Chapter 8 - Predominantly Waterfowl									
Anatipestifer disease, New Duck Syndrome, Duck Septicaemia	R	X	X			R	R		241
Coccidiosis kidney				X					242
Coccidiosis, Intestinal, of Ducks and Geese			X	X					243
Duck viral hepatitis			X						244
Duck virus enteritis, duck plague			X	X					245
Gizzard worms - Geese			X	X					246
Goose Parvovirus (Derzsy's Disease)			X	X					247
Mycoplasma immitans infection			X	X					248
Pseudotuberculosis	X	X	X	X	X	X			248
Streptococcus bovis septicaemia			X						249

DISEASES AND SYNDROMES: CHICKENS AND VARIOUS SPECIES

CHAPTER 6

DISEASES AND SYNDROMES: CHICKENS AND VARIOUS SPECIES

AMYLOIDOSIS

Introduction
A disease of ducks and chickens caused by prolonged stimulation of the immune system usually due to chronic bacterial infection e.g. enterococcal or staphylococcal arthritis and perhaps *Mycoplasma synoviae*. Common in adult parent ducks, it also occurs in laying- and meat-type chickens. The disease itself is non-transmissible, though the predisposing primary infection may be. Morbidity is usually low, the course of the disease chronic and mortality is high.

Signs
    Wasting.
    Lameness.
    Swelling of foot and leg joints.

Post-mortem lesions
    Enlarged liver, parboiled in appearance.
    Spleen enlarged, haemorrhagic or ruptured.
    Orange-coloured deposits in joints.
    Primary inflammatory lesions (arthritis, endocarditis etc).

Diagnosis
    Lesions, histopathology. Differentiate from diffuse tumours of the liver.

Treatment
    None.

Prevention
    Control of predisposing infection.

Ascites, 'Water-Belly', Broiler Ascites Syndrome

Introduction
Associated with inadequate supplies of oxygen, poor ventilation and physiology (oxygen demand, may be related to type of stock and strain). Ascites is a disease of broiler chickens occurring worldwide but especially at high altitude. The disease has a complex aetiology and is predisposed by reduced ventilation, high altitude, and respiratory disease. Morbidity is usually 1–5%, mortality 1–2% but can be 30% at high altitude. Pulmonary arterial vasoconstriction appears to be the main mechanism of the condition.

Signs
Sudden deaths in rapidly developing birds.
Poor development.
Progressive weakness and abdominal distension.
Recumbency.
Dyspnoea.
Possibly cyanosis.

Post-mortem lesions
Thickening of right-side myocardium.
Dilation of the ventricle.
Thickening of atrioventricular valve.
General venous congestion.
Severe muscle congestion.
Lungs and intestines congested.
Liver enlargement.
Spleen small.
Ascites.
Pericardial effusion.
Microscopic – cartilage nodules increased in lung.

Diagnosis
Gross pathology is characteristic. A cardiac specific protein (Troponin T) may be measured in the blood. This may offer the ability to identify genetic predisposition. Differentiate from broiler Sudden Death Syndrome and bacterial endocarditis.

Treatment
Improve ventilation, Vitamin C (500 ppm) has been reported to be of benefit in South America.

Prevention
Good ventilation (including in incubation and chick transport), avoid any genetic tendency, control respiratory disease.

ASPERGILLOSIS

Introduction
A fungal infectious disease, caused by *Aspergillus fumigatus*, in which the typical sign is gasping for breath, especially in young chicks. Sometimes the same organism causes eye lesions or chronic lesions in older birds. The fungus can infect plant material and many species of animals including birds and man. Occasionally similar lesions are produced by other species of *Aspergillus* or even other fungi such as *Penicillium*, *Absidia* etc.

It affects chickens, turkeys, ducks, penguins, game birds, waterfowl, etc, worldwide. The infection has an incubation period of 2–5 days. Morbidity is usually low, but may be as high as 12%. Mortality among young affected birds is 5–50%. Transmission is by inhalation exposure to an environment with a high spore count; there is usually little bird-to-bird transmission. Spores are highly resistant to disinfectants.

Signs
Acute form:
Inappetance.
Weakness.
Silent gasping.
Rapid breathing.
Thirst.
Drowsiness.
Nervous signs (rare).

Chronic Forms:
Ocular discharge (ocular form only).
Wasting.

Post-mortem lesions
Yellow to grey nodules or plaques in lungs, air sacs, trachea, plaques in peritoneal cavity, may have greenish surface.
Conjunctivitis/keratitis.
Brain lesions may be seen in some birds with nervous signs.

Diagnosis
This is usually based on the signs and lesions and microscopic examination for the fungus, preferably after digestion in 10% potassium hydroxide. It may be confirmed by isolation of the fungus, typically by putting small pieces of affected tissue on Sabouraud agar. Growth occurs in 24–48 hours and colonies are powdery green/blue in appearance. Differentiate from excessive exposure to formalin or vaccinal reactions in day olds and from heat stress in older birds.

Treatment
Usually none. Environmental spraying with effective antifungal antiseptic may help reduce challenge. Amphotericin B and Nystatin have been used in high-value birds.

Prevention
Dry, good quality litter and feed, hygiene, Thiabendazole or Nystatin has been used in feed.

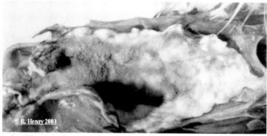

Figure 8. 'Furry' airsacculitis in aspergillosis of an adult duck. The powdery surface is dark green in colour.

AVIAN ENCEPHALOMYELITIS EGG DROP

Introduction
Avian encephalomyelitis virus infection in laying bird causes

inapparent infection or drops in egg production. See 'Avian Encephalomyelitis, Epidemic tremors' for its effect in young birds. It affects chickens, turkeys, quail, pheasants and occurs in most poultry-producing countries. Morbidity 5–60%, mortality none. The means of transmission is unknown but probably by faecal contamination of environment, feed, water etc. with an oral infection route. Virus in faeces may survive 4 weeks or more. Predisposed by immunosuppression.

Signs
Drop in egg production, small (5–10%) and lasting no more than 2 weeks.
In breeders there may be a drop in hatchability of about 5%, and there is serious disease in the progeny (see next section).

Post-mortem lesions
None.

Diagnosis
History, rising titre to AE virus, subsequent disease in progeny if breeders. Serology – The embryo protection test has been used in the past, now Elisa is used more commonly. Differentiate from Infectious Bronchitis, lentogenic Newcastle disease, EDS-76.

Treatment
None.

Prevention
Vaccination of breeders/layers at 9–15 weeks, attenuated or not. Immunity is usually long lasting.

AVIAN ENCEPHALOMYELITIS, EPIDEMIC TREMORS

Introduction
Avian encephalomyelitis is a viral disease of the central nervous system of chickens, pheasants, turkeys, and quail. It has a worldwide distribution. Morbidity 5–60% depending on the immune status of the majority of parents, mortality high. Vertical transmission is very important, transmission occurs over about 1–2 weeks, some

lateral. The route of infection is transovarian with an incubation period of 1–7 days; lateral transmission is probably by the oral route, incubation >10 days. Virus in faeces may survive 4 weeks or more.

Signs
Nervous signs.
Dull expression.
Ataxia and sitting on hocks.
Imbalance.
Paralysis.
Tremor of head, neck and wings. Tremor may be inapparent but is accentuated if chicks are held inverted in the hand.

Post-mortem lesions
Gross lesions are mild or absent.
There may be focal white areas in gizzard muscle (inconstant). A few recovered birds may develop cataracts weeks after infection.
Microscopic – nonpurulent diffuse encephalomyelitis with perivascular cuffing.

Diagnosis
A presumptive diagnosis is based on the history, signs, and lack of significant lesions. Histopathology is usually diagnostic and IFA, and/or viral isolation may be carried out if required. The embryo protection test has been used in the past, now Elisa is used more commonly. Differentiate from Newcastle disease, vitamin deficiency (E, A, riboflavin), toxicities, EE (especially in pheasant in the Americas), Marek's disease, Mycotic Encephalitis, Brain abscess, *Enterococcus hirae* infection.

Treatment
None.

Prevention
Vaccination of breeders at 9–15 weeks, attenuated or not. Immunity is long lasting.

Avian Leukosis, Lymphoid Leukosis, Leukosis/Sarcoma group

Introduction

A complex of viral diseases with various manifestations such as lymphoid leukosis, myeloblastosis (see below for Sero-type J), erythroblastosis, osteopetrosis, myxosarcomas, fibrosarcomas, other tumours. It affects chickens worldwide with susceptibility varying considerably among different strains and types of stock – egg layers are generally more susceptible to lymphoid leukosis. Morbidity is low but mortality high. Mortality tends to be chronically higher than normal for a prolonged period. Egg production is somewhat reduced. There may be increased susceptibility to other infectious diseases due to damage to the immune system. Vertical transmission is most important by infection of the egg white in infected breeders (who are long-term carriers), lateral transmission is poor but infection may occur by the faecal–oral route, especially in young birds. In lymphoid leukosis the incubation period is about 4–6 months; it may be as short as 6 weeks for some of the other manifestations. The causative viruses are rapidly inactivated at ambient temperature and on exposure to most disinfectants.

Signs
Depression.
Emaciation.
Loss of weight.
Persistent low mortality.
Enlargement of abdomen, liver or bursa.
Many are asymptomatic.

Post-mortem lesions
Focal grey to white tumours, initially in the bursa, then liver, spleen, kidney etc. Liver may be very large.
Microscopic – cells lymphoplastic

Diagnosis
History, age, lesions, cytology. Differentiate from Marek's disease, coligranuloma.

Treatment
None.

Prevention
Good hygiene, all-in/all-out production, control arthropods, eradication – checking of antigen in the albumen is a basis for eradication (see Sero-type J for details).

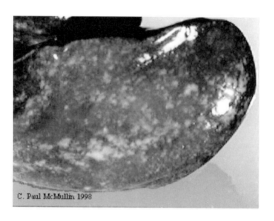

Figure 9. Diffuse lymphoid tumours in an enlarged liver from a mature broiler parent hen. This was a case of Myelocytoma Avian Leukosis (Sero-type J).

AVIAN LEUKOSIS (SERO-TYPE J), MYELOCYTOMATOSIS

Introduction
Caused by an avian retrovirus. This condition has until now been seen only in meat-type chickens, with considerable strain-to-strain variation. It has occured in Europe, North and South America. Morbidity is low, though there is high mortality of affected birds. Transmission is by congenital infection from antibody-negative females, bleeding and vaccination needles; lateral transmission by faecal–oral route (this declines as the bird ages). The incubation period is 10–20 weeks. Congenitally infected birds tend to remain antibody negative, shed virus and develop tumours. Virus survival

is poor but sufficient to allow cross contamination in hatcheries and on farm in rear.

Signs
Depression.
Emaciation.
Loss of weight.
Persistent low mortality.
Enlargement of abdomen, liver.
Many are asymptomatic.

Post-mortem lesions
Liver enlargement, often with tumour foci.
Splenomegaly and enlarged kidneys also occur.
Most characteristically are chalky white tumours in the bone marrow, particularly of the sternum, ribs, sacral joint.
Microscopic – tumours usually contain well-differentiated myelocytes. Two cell types may be found in the same tumour.

Diagnosis
History, age, lesions, histology, ultimately identification of virus by isolation and/or PCR. Differentiate from Marek's disease, Lymphoid Leukosis.

Treatment
None.

Prevention
Checking of antigen in the albumen is a basis for eradication – most but not all, birds with egg antigen will be antibody negative. 'Shedders' – 80% produce infected chicks, 'non-shedders' – only 3% produced infected chicks. PCR testing of embryonally infected chicks using DNA testing is uniformly positive for blood and faecal samples. There is also evidence that it is slightly more sensitive than conventional testing. Serology – an Elisa test is aivailable to identify antibody positive birds. Prevent/reduce cross-infection in hatchery and on farm.

Critical hatchery practices:
Separate infected and uninfected lines.
Handle clean lines before infected lines, preferably on separate

hatch days and in separate machines.
Separation in vaccination.
Minimise stress.

Farm practices:
Brood and rear lines separately and maintain separate for as long as possible.
Minimise group sizes.
Delay live vaccine challenges.
Avoid migration errors (birds unintentionally moving between pens).

AVIAN RHINOTRACHEITIS, 'SWOLLEN HEAD SYNDROME'

Introduction
A viral disease of chickens, turkeys (see separate summary), guinea fowl and possibly pheasants seen in Europe, Africa, South America and North America. It is caused by a pneumovirus of the Paramyxoviridae family, first isolated from poults in South Africa in 1978. Two subgroups have been identified on the basis of the G-protein sequence: A (original UK isolates) and B (original southern Europe isolates). There is rapid lateral transmission with infection by aerosol through the respiratory route; vertical transmission is uncertain. As for many infections, fomites can be important in moving infection between farms. The incubation period is 5–7 days, morbidity is 10–100% and mortality can be 1–10%.

Signs
Decreased appetite, weight gain and feed efficiency.
Facial and head swelling (though this can occur in other conditions).
Loss of voice.
Ocular and nasal discharge.
Conjunctivitis.
Snick.
Dyspnoea.
Sinusitis.

Post-mortem lesions

Serous rhinitis and tracheitis, sometimes pus in bronchi.

If secondary invasion by *E. coli* then pneumonia, airsacculitis and perihepatitis.

Congestion, oedema and pus in the air space of the skull occurs in a proportion of affected birds due to secondary bacterial infections

Diagnosis

Clinical signs, serology, isolation of ciliostatic agent. Differentiate from Infectious Bronchitis, Lentogenic Newcastle disease, low virulence avian influenza, *Ornithobacterium rhinotracheale*. Serology – Elisa normally used, not all commercial kits are equally sensitive to response to both A and B challenge viruses.

Treatment

Antibiotic not very effective. Control respiratory stressors, chlorination of drinking water, multivitamins.

Prevention

All-in/all-out production, vaccination (degree of cross protection between A and B types remains to be established). Live vaccines can reduce clinical signs and adverse effects, inactivated vaccines may be used in breeders prior to lay.

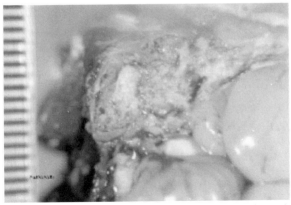

Figure 10. Pus in skull bones. This is a common sequel to avian pneumovirus infection in both chickens and turkeys.

Beak Necrosis

Introduction
A condition seen in chickens and turkeys caused by excessively fine mashed feeds.

Signs
Feed accumulates along edges of lower beak leading to distortion and necrosis of horny tissue.

Has been associated with curled tongue in turkeys in the past. This may also be associated with ulcers in the roof of the mouth (palate) which can be confused with some forms of mycotoxicosis.

Post-mortem lesions
See signs.

Diagnosis
Clinical signs, exclusion of other causes of similar signs.

Treatment
Not usually sufficiently severe to justify medication. Mild water sanitation may help control secondary infections in the affected tissues.

Prevention
Check feed particle size by granulometry, grind less finely.

Bedbug Infestation

Introduction
A condition of poultry, pigeons (and mammals) caused by infestation with the external parasite *Cimex lectularius*. The parasites are up to 5 mm long and feed at night. It occurs mainly in subtropical and some temperate areas. Adult parasites can survive for 1–12 months in the environment without feeding. Eggs laid by the adult parasites hatch in 4–20 days, there are five nymphal stages each of which feed on birds. Growth to adult parasite takes 1–3 months, depending on temperature.

Signs
Lack of thrift.
Anaemia.
Reduced production when infestation is serious.

Post-mortem lesions
Anaemia.

Diagnosis
Identification of the parasite. Differentiate from other blood-sucking parasites.

Treatment
Appropriate insecticide treatment of the environment, in particular the cracks and crevices where the parasites hide during the daytime.

Prevention
Thorough treatment of the empty building at turn-around with an appropriate insecticide. Fumigation is also helpful.

BIG LIVER AND SPLEEN DISEASE

Introduction
This condition was first seen in Australia in 1980. Although the exact cause has not been identified it is believed to be caused by a virus and the lesions are associated with deposition of antigen/antibody complexes in tissues. Only chickens are known to be affected, most commonly broiler parents in lay. Natural infections have only been demonstrated in birds over 24 weeks of age though it is possible that vertical transmission and/or infection in rear occur with a subsequent period of latency. Embryos inoculated intravenously become persistently antigen positive but only show antibody much later.

Signs
Chronic under production or egg drops of up to 20%.
Mortality of up to 1% per week for 3–4 weeks.
Anaemia.
Premature moulting.

Post-mortem lesions

Enlarged spleen (over 1gm/kg bodyweight, often with pale foci). Liver usually enlarged, sometimes with subcapsular haemorrhage.

Affected birds may also have:

Lung congestion
Ovarian regression
Yolk peritonitis
Pale foci and haemorrhages in pancreas.

Diagnosis

Typical signs and lesions. May be confirmed by serological tests to detect either the specific antigen (which is likely to be positive first) or the response to it.

Treatment
None known.

Prevention

Thorough cleaning and disinfection after depletion of an affected flock. Good biosecurity. All-in/all-out production.

BIOTIN DEFICIENCY, INCLUDING FATTY LIVER AND KIDNEY SYNDROME

Introduction

Biotin deficiency has occured in turkeys and chickens in many countries but is now rare in birds consuming properly formulated feeds. Reduced feed intake and blood sugar can precipitate fatty liver and kidney syndrome.

Signs

Poor growth.
Leg weakness.
Scabs around eyes and beak.
Thickened skin under foot pad, in embryos, webbing between toes.

Chondrodystrophy.
Sudden deaths in fatty liver and kidney syndrome.

Post-mortem lesions
See signs.
Pale livers and kidneys in fatty liver and kidney syndrome.

Diagnosis
Signs, lesions, response to treatment/prevention. Differentiate from pantothenic acid deficiency (skin lesions).

Treatment
Addition of biotin in feed or water.

Prevention
Supplementation of diets with biotin – naturally present in many raw materials, has very low bioavailability.

BITING LICE

Introduction
Various species of lice are common external parasites of poultry worldwide. They are spread by direct contact between birds and by litter etc. Away from birds adults survive about 4–5 days. The parasites are 1–6 mm in size and their life cycle takes about 3 weeks. *Menocanthus stramineus* is the most pathogenic and is said to be capable of causing anaemia in heavy infestations. Check flocks regulary for rapidly moving insects at the base of the feathers on the abdomen or around the vent. Crusty clumps of eggs ('nits') may be visible at the base of feathers.

Signs
Lack of thrift in young birds.
Lice eggs stuck to feathers.
Parasites on birds, especially around vent.
Irritation.
Loss of vent feathers.
Scabs around vent.
Loss of condition.

Drop in egg production.

Post-mortem lesions
Usually none, may be some feather damage and crustiness of skin.

Diagnosis
Identification of the parasites. Differentiate from mites, bedbugs.

Treatment
Malathion powders and pyrethroid sprays where approved for bird application.

Prevention
Avoid direct contact with wild and backyard poultry. Examine for lice regularly, especially in autumn and winter and treat if required. It is usually necessary to treat twice at a 7–10 day interval to fully control the condition, as the larvae within eggs are not killed by most products. Effective removal of all organic material at flock depletion should be practised in all-in/all-out production systems.

BLACKFLY INFESTATION

Introduction
Grey-black hump-backed flies, 5 mm long and found in North and South America that are external parasites of birds and mammals. The flies transmit leucocytozoonosis and also a filarial parasite in ducks. The condition tends to occur near to rapidly flowing streams, although these insects can travel up to 15 miles. Eggs and larvae survive through the winter to cause new infestations in the following year.

Signs
Anaemia in young birds.
Swarms of flies.

Post-mortem lesions
Anaemia.

Diagnosis
Anaemia, season, local history.

Treatment
Treatment is difficult.

Prevention
Similar measures as for mosquito control. Biological control using a strain of *Bacillus thuringensis* has had some success and is preferable to insecticides. Weekly treatments are required.

BOTULISM

Introduction
A condition of chickens, turkeys, ducks and other waterfowl occurring worldwide and caused by a bacterial toxin produced by *Clostridium botulinum* mainly types A / C. The toxin is produced in decaying animal (usually carcases) and plant waste, and toxin-containing material (pond-mud, carcases, maggots) is consumed by the birds. Toxin may also be produced by the bacteria in the caecum. Morbidity is usually low but mortality is high. The toxin and bacterial spores are relatively stable and may survive for some time in the environment. It has also been suggested that poultry carcases lost in litter can be a cause of botulism in cattle grazing land or consuming silage where poultry litter has been spread.

Signs
Nervous signs, weakness, progressive flaccid paralysis of legs, wings then neck, then sudden death.
Affected broilers tend to settle with eyes closed when not disturbed.
A soiled beak, because it rests on the litter, is also quite typical.

Post-mortem lesions
Possibly no significant lesions.
Mild enteritis if has been affected for some time.
Feathers may be easily pulled (chicken only).
Maggots or putrid ingesta may be found in the crop.

Diagnosis
History, signs, mouse toxicology on serum or extract of intestinal contents. Differentiate from acute Marek's disease ('Floppy Broiler Syndrome') by histology of the brain.

Treatment
Remove source of toxin, supportive treatment if justifiable, antibiotics, selenium.

Prevention
Preventing access to toxin, suspect food and stagnant ponds, especially in hot weather. The single most important measure is careful pick-up and removal of all dead birds on a daily basis. This will reduce the risk of botulism both in the poultry and in any grazing animals on land where poultry litter is spread.

BREAST BLISTER

Introduction
A complex condition of chickens and turkeys occurring worldwide associated with trauma, leg weakness, and infection with *Staphylococcus* spp. bacteria. Morbidity may reach more than 50% but the condition is not fatal. Poor feather cover and caked or wet litter are predisposing factors.

Signs
Swelling over the keel bone with bruising and discolouration.

Post-mortem lesions
Inflammation of sternal bursa along the keel bone which may, in chronic cases, give way to scar tissue.

Diagnosis
Based on lesions.

Treatment
Not usually appropriate.

Prevention
Good litter management and handling, control of leg problems.

Chapter 6 - Diseases and Syndromes: Chickens & Various Species

CAECAL WORM

Introduction
Heterakis gallinae, nematode parasites of poultry and game birds, are small whitish worms with a pointed tail, up to 1.5 cm in length that occur in the caecum. They are found worldwide. Morbidity is high but it is not associated with mortality. Infection is by the oral route. Earthworms may be transport hosts for eggs, or paratenic hosts with partially developed (L2) larvae. There is an incubation period of 2 weeks for eggs to embryonate, and a four-week prepatent period. The meaning of the technical terms relating to parasite life cycle are defined in the glossary. *Heterakis gallinae* eggs and larvae are a transport hosts for *Histomonas*, the cause of Blackhead.

Signs
None.

Post-mortem lesions
Inflammation of caecum, possibly with nodule formation.

Diagnosis
Adults can be seen in caecal contents at post-mortem examination.

Treatment
Flubendazole, Levamisole, are effective.

Prevention
Avoiding access to earth and earthworms. Routine anthelmintic treatment.

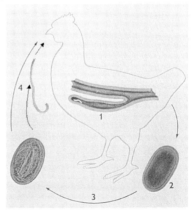

Figure 11. The life cycle of *Heterakis* showing the location of adults, an undeveloped egg as found in fresh faeces, and the embryonated egg. The egg may be ingested directly by a chicken, or by an earthworm which is in turn ingested by a chicken.

CALCIUM TETANY

Introduction
A metabolic disease of chickens, especially broiler parents. Predisposing factors include heat stress with reduced feed intake and panting.

Signs
Paralysis.
Death from respiratory and cardiac failure.

Post-mortem lesions
Cyanosis.
Congested lungs.
Active ovary with egg in oviduct.

Diagnosis
This is made on signs, lesions, lack of other significant lesions, and response to treatment. Differentiate from IB 793b, other acute infections and other causes of sudden death.

Treatment
Provide 5 gm of oyster shell per hen on 3 successive days along with vitamin D in drinking water.

Prevention
Keep pullet flocks on low calcium diet until 5% production hen/day, managing birds for maximum uniformity.

CAMPYLOBACTER INFECTION

Introduction
Campylobacter spp. are bacteria that commonly infect a broad range of livestock species, pets and wild animals. In poultry they tend to multiply in large numbers in the hindgut, principally in the caecae. Campylobacters are a significant cause of enteritis in man. Infected poultry are a potential reservoir of this zoonosis. *Campylobacter jejuni* is the commonest species found in poultry. All campylobacters are delicate organisms that survive for relatively

short periods outside the host unless protected by organic material, biofilm or engulfed by protozoa. *Campylobacter jejuni* infection is not currently considered to be pathogenic in poultry though a *Campylobacter*-like organism is considered to be the cause of 'Vibrionic Hepatitis'. There are indications that plantar pododermatitis, carcase quality and litter quality are better on farms which tend to have *Campylobacter*-negative stock. The reason for this is unclear. It may be that management that favours dry litter reduces the risk of infection and/or transmission within the flock. There is an annual cycle with increased risk of infection in the summer months in some countries.

Signs
None.

Post-mortem lesions
None.

Diagnosis
Isolation of the organism from caecal contents, cloacal swabs or composite faeces. The organism is sensitive to air so swabs should be collected into transport medium and other samples placed in airtight containers with minimal airspace. Samples should be tested as quickly as possible after collection.

Treatment
Not required on clinical grounds.

Prevention
In principle, housed poultry can be maintained free of *Campylobacter* infection by consistent application of excellent biosecurity. Key aspects of this include effective sanitation of drinking water, sourcing of water from high quality supplies, avoidance of contact with pets and other farmed species, good hand hygiene by stockmen, and changing of overalls and boots on entering bird areas. In practice the success of this will also depend upon the degree of environmental contamination by the organism. For this reason it may be difficult to stop the spread of infection between houses once it becomes established in one house. Many infections are introduced during thinning or other forms of partial depopulation. Insects and rodents may act as a

means of transfer of the infection from the general environment into the poultry buildings. Research is ongoing on the development of vaccines, phage treatments and competitive exclusion approaches, as well as processing plant technologies to reduce carcase contamination.

CANDIDIASIS, MONILIASIS, THRUSH

Introduction

A disease of the alimentary tract of chickens, turkeys, and sometimes other birds and mammals, characterised by thickening and white plaques on the mucosa, especially in the crop but sometimes in the proventriculus, intestine and cloaca, and associated with gizzard erosion. The cause is a fungal yeast, *Candida albicans* and the condition is seen worldwide. Morbidity and mortality are usually low. The route of infection is normally oral and the organism is often present in healthy animals with disease occurring secondary to stress and poor hygiene. The fungus is resistant to many disinfectants.

Signs

Dejection.
Poor appetite.
Slow growth.
Diarrhoea, possibly confused or masked by signs of the primary disease.

Post-mortem lesions

White plaques in mouth, oesophagus, crop, occasionally proventriculus and intestine.
Raised focal lesions may slough into lumen as caseous material.

Diagnosis

Lesions, histopathology, microscopic examination of a digested smear (heat in 10% potassium hydroxide) to demonstrate the hyphal forms of the yeast in the tissues. Colonies of this fungus appear as white to ivory colour, smooth and with a yeasty smell.

Treatment
Nystatin (100 ppm in feed) for 7–10 days, copper sulphate (1 kg/tonne feed) for 5 days, or copper sulphate 1gm/2 litre water for 3 days if approved locally.

Prevention
Avoid excessive use of antibiotics and other stressors. Ensure good hygiene, proprionic acid, sodium or calcium proprionate at 1 kg per tonne continually. A finely divided powder of copper sulphate (where approved) at 200gm/tonne continually or to 14–16 weeks in replacement pullets. Control of *Candida* through drinking water is sometimes practised with chlorination (e.g. Chlorox, sodium hypochlorite) at 5 ppm. This is economical and effective. It should be repeated periodically. Take care to provide fresh clean feed and water, uncontaminated by fungi.

CANNIBALISM, FEATHER PECKING

Introduction
A complex multifactorial behavioural problem of poultry and game birds seen worldwide. Morbidity is usually low but mortality is high among affected birds. Predisposing factors include overcrowding, excessive light intensity or variation (e.g. through shafts of light in the house), high temperatures, nutritional deficiencies, feed form (mash takes longer to consume than pellets), tenosynovitis and other diseases affecting mobility, boredom, and strain of bird.

Signs
Pecking at feet (especially young chicks) and vents (adult layers and turkey poults 8–12 days old), head, face, wings.
Feather-pulling.

Post-mortem lesions
Skin wounding related to particular signs exhibited.
Generalised anaemia.

Diagnosis
Age, distribution of lesions, anaemia. Differentiate from bacterial dermatitis, post-mortem cannibalism.

Treatment
Correct any husbandry problems. Soluble multivitamins and/or methionine may be of some benefit in some circumstances. Beak trimming may be necessary. If so it should be carried out carefully by trained operators, complying with local regulations and any relevant codes of practice.

Prevention
Proper density and temperature, low light level, control ectoparasites. Provision of a diet that closely matches the nutritional requirements of the stock concerned.

CAPILLARIASIS – HAIRWORM INFECTION

Introduction
Nematode parasitic worms of poultry, game birds and pigeons of *Capillaria* species. *C. obsignata* in the small intestine, *C. contorta* in the crop and oesophagus. The worms are 7–18 mm long, about 0.05 mm wide and hair-like in appearance. Morbidity and mortality are usually low. Infection is by the oral route. Worm eggs take about 20 days to embryonate with an L1 larvae, prepatent period about 21–25 days according to species. Some species have earthworms as intermediate hosts; some are transmitted direct from bird to bird. Worm eggs in the environment are resistant.

Signs
Diarrhoea.
Wasting
Poor growth.
Dejection.

Post-mortem lesions
Enteritis.
Hairworms in mucosa of crop, small intestine or caecum.

Diagnosis
This may be by a combination of macroscopic examination, seiving intestinal contents, or characteristic worm eggs in faeces in patent infections. Differentiate from other causes of enteritis.

Treatment
Coumphos has been licensed in some markets. Fenbendazole has been shown to have high efficacy – other approved benzimidazoles can be expected also to have activity. Levamisole.

Prevention
Separation of birds from possible transport and intermediate hosts, effective cleaning of houses.

CELLULITIS

Introduction
Cellulitis is literally an inflammation of connective tissues. It typically occurs between skin and muscles and between muscles and may be an incidental finding in a range of conditions. However its main importance is as a cause of condemnation in meat poultry, particularly broiler chickens. In the USA it is called 'Inflammatory Process'. The condition is caused by infection of, often minor, skin wounds by particular strains of *E. coli*, which can replicate in the tissues.

Signs
Affected flocks tend to have poorer than average productivity and uniformity, but the affected birds are not readily detectable prior to slaughter.

Post-mortem lesions
Typically it presents as exudate ranging from liquid and pale cream pus to yellowish solid plaques of caseous material under the skin of the abdomen and/or in the leg. Many affected birds have no other lesions and are reasonably well grown. Many meat inspectors become skilled at detecting subtle differences in skin colour in the affected birds.

Diagnosis
Typical lesions.

Treatment
Treatment would not be possible if the problem is identified at

a final depletion. If identified at a thinning there may be time for antibacterial treatment to have some benefit for those birds in the early stages of the problem.

Prevention

Toe scrapes at 15–25 days of age when feather cover is poor are the most likely predisposing factors. Careful flock management with a view to reducing toe wounds has the greatest impact in controlling cellulitis. Routine monitoring of skin damage at about 25 days of age may be helpful in fostering good practices, though most of the birds showing toe scrapes will not go on to develop cellulitis.

CHICKEN ANAEMIA

Introduction

A viral disease of chickens caused by Chicken Anaemia Virus or CAV. Prior to confirmation that it is in fact a virus it was known as Chicken Anaemia Agent or CAA. Mortality is typically 5–10% but may be up to 60% if there are predisposing factors present such as intercurrent disease (Aspergillosis, Gumboro, Inclusion body heptatitis etc.) or poor management (e.g. poor litter quality). Transmission is usually vertical during sero-conversion of a flock in lay, lateral transmission may result in poor productivity in broilers. The virus is resistant to pH 2, ether, chloroform, heat (70°C for 1 hour, 80°C for 5 minutes) and many disinfectants even for 2 hours at 37°C. Hypochlorite appears most effective *in vitro*.

Signs

Poor growth.
Pale birds.
Sudden rise in mortality (usually at 13–16 days of age). No clinical signs or effect on egg production or fertility in parent flock during sero-conversion.

Post-mortem lesions

Pale bone marrow.
PCV of 5–15% (normal 27–36%).
Atrophy of thymus and bursa.

Discoloured liver and kidney.
Gangrenous dermatitis on feet, legs wings or neck.
Acute mycotic pneumonia.

Diagnosis
Gross lesions, demonstration of ongoing sero-conversion in parent flock, virus may be isolated in lymphoblastoid cell line (MDCC-MSB1).

Treatment
Good hygiene and management, and control of other diseases as appropriate, may be beneficial. If gangrenous dermatitis is a problem then periodic medication may be required.

Prevention
Live vaccines are available for parents, their degree of attenuation is variable. They should be used at least 6 weeks prior to collecting eggs for incubation. Their use may be restricted to those flocks that have not sero-converted by, say, 15 weeks. Immunity: there is a good response to field challenge (in birds over 4 weeks of age) and to attenuated live vaccines. Serology: antibodies develop 3–6 weeks after infection, and may be detected by SN, Elisa, or IFA.

CHONDRODYSTROPHY, SLIPPED TENDON OR PEROSIS

Introduction
Caused by deficiency of manganese, choline, zinc, either singly or in combination (although deficiencies of pyridoxine, biotin, folic acid, niacin may also be involved). This condition is seen in chickens, ducks and turkeys. In turkeys it may be an inherited deficiency of galactosamine.

Signs
Short legs.
Lameness.
Distortion of hock.
Slipping of Achilles tendon (or perosis).

Malposition of leg distal to hock.
In embryos parrot beak, shortened bones.

Post-mortem lesions
Shortening and thickening of long bones.
Tibia and metatarsus bowed.
Shallow trochlea.
Lateral slipping of tendon.

Diagnosis
Lesions, analysis of feed. Differentiate from twisted leg, infectious synovitis, rickets, infectious arthritis, ruptured ligaments.

Treatment
For flock proceed as for prevention, no value to affected bird.

Prevention
Addition of manganese, choline, vitamins, correct mineral balance.

COCCIDIOSIS, UPPER INTESTINAL, *E. ACERVULINA*

Introduction
This is probably the commonest cause of coccidiosis in chickens and occurs worldwide. It is seen in layers and in broilers, both alone and in association with other species of coccidia and is caused by *Eimeria acervulina*, which is moderately pathogenic. Morbidity is variable and mortality low or absent. *Eimeria mivati* is currently considered not to be a valid species distinct from *E. acervulina*.

Signs
Depression.
Ruffled feathers.
Closed eyes.
Inappetance.
Poor production.
Diarrhoea.
Depigmentation.

Post-mortem lesions

Thickening, and other lesions, restricted to upper third of small intestine – the duodenum and part of the ileum.

Petechiae.

White spots or bands in the mucosa. In severe infections they become confluent and cause sloughing of the mucosa.

Poor absorption of nutrients/pigments.

A system of assessing the severity of coccidial challenge by attributing a 'score' is often used. A detailed description is beyond the scope of this book. In general terms a score of 0 indicates no lesions and a score of 4 indicates maximal severity of lesion or death. Various publications provide a photographic key to severity of lesion.

Diagnosis

Signs, lesions, microscopic exam of scrapings. Differentiate from necrotic and non-specific enteritis.

Treatment

Toltrazuril, Sulphonamides, Amprolium, in feed or water.

Prevention

Coccidiostats in feed, vaccination by controlled exposure, hygiene. Immunity is quite short lived (about 30 days) in the absence of continued challenge.

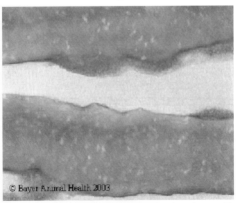

Figure 12. Moderate *Eimeria acervulina* infection (score 2) in chicken duodenum. In milder infections there may be scattered white spots, in severe the entire surface is pale or denuded of epithelium.

Coccidiosis, Mid-intestinal, *E. maxima*

Introduction
One of the more common forms of coccidiosis in commercial broilers. Because of the area of intestine affected it tends to have a significant effect on productivity and susceptibility to necrotic enteritis. Caused by *Eimeria maxima*, of moderate to high pathogenicity it is seen worldwide. Morbidity and mortality are variable.

Signs
Depression.
Ruffled feathers.
Closed eyes.
Inappetance.
Poor production.
Blood or pigment in the faeces.
Depigmentation of skin and plasma is especially evident in this form of cocccidiosis and this is commercially important in some markets.

Post-mortem lesions
Petechiae and thickening of middle third of intestine.
Poor absorption of nutrients/pigments.
Mild to severe enteritis, contents often orange in colour, mucosa tends to be pinker than normal.
This infection is often associated with *E. acervulina* coccidiosis and there may be large numbers of characteristic oocysts in smears.

Diagnosis
Signs, lesions, microscopic examination of scrapings. Differentiate from necrotic enteritis, non-specific enteritis.

Treatment
Sulphonamides, Amprolium, Vitamins A and K in feed or water.

Prevention
Coccidiostats in feed, vaccination, hygiene. This is one of the less immunogenic species, commercial vaccines commonly contain more than one strain of *E. maxima*.

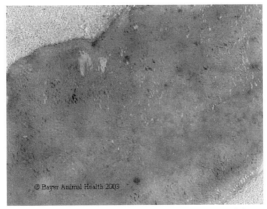

Figure 13. Moderate *Eimeria maxima* infection in the jejunum of a chicken (Score 2). The lesions are subtle compared to other forms of coccidiosis. The intestine is slightly thickened and there are scattered haemorrhages in the mucosa when seen from the inside.

COCCIDIOSIS, MID-INTESTINAL, *E. NECATRIX*

Introduction
A highly pathogenic form of coccidiosis, caused by *Eimeria necatrix*, in which the parasite is present in the small intestine and in the caecum. It occurs in chickens worldwide and has variable morbidity but mortality is high in severely affected birds.

Signs
Reduced feed consumption.
Depression.
Ruffled feathers.
Closed eyes.
Inappetance.
Poor production.
Diarrhoea, blood in faeces.

Post-mortem lesions
Petechiae and thickening, of middle to posterior third or more of small intestine.

'Sausage-like' intestine.
Severe necrotising enteritis.
Schizonts seen as white spots through the serosa interspersed with petechiae. Deep scrapings necessary to show large schizonts. Oocyts in caecal scrapings.

Diagnosis
Signs, lesions, microscopic examination of scrapings
Differentiate from necrotic enteritis, other types of coccidiosis.

Treatment
Toltrazuril, Sulphonamides, Amprolium, Vitamins A and K in feed or water.

Prevention
Coccidiostats in feed, vaccination, hygiene. This is one of the less immunogenic species, commercial vaccines commonly contain more than one strain of *E. maxima*.

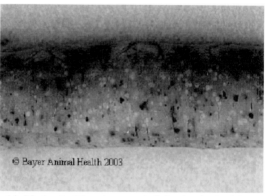

Figure 14. Moderate *Eimeria necatrix* infection in the jejunum of a chicken (Score 3). In this case the intestine is thickened and can become ballooned and sausage-like. Haemorrhages and white spots are visible from the outside of the intestine.

COCCIDIOSIS, *E. MITIS*

Introduction
This condition of chickens, seen worldwide, is caused by the

protozoan parasite *Eimeria mitis*, which colonises the small intestine. The infective agent is found in litter, faeces and on fomites and birds are infected by the oral route with an incubation period of 2–5 days. The disease occurring is proportional to the amount of infective agent ingested. The parasite is moderately resistant in the environment and highly resistant to conventional disinfectants. Predisposing factors include exposure to faeces and litter conditions that favour development of the parasite (temperature, humidity).

Signs
Reduced feed conversion efficiency and weight gain. May predispose to wet litter, secondary bacterial enteritis.

Post-mortem lesions
The lesions are minimal and located in the lower small intestine (ileum) which tends to be pale and flaccid with scattered petechiae.

Diagnosis
Mild lesions, identification of typical small round oocysts and other stages in fresh scrapings from the small intestine.

Treatment
Not usually treated but susceptible to the products used for other forms of intestinal coccidiosis.

Prevention
Normally controlled by anticoccidials in feed. May be included in vaccines.

COCCIDIOSIS, *E. PRAECOX*

Introduction
Infection of chickens with the protozoan parasite *Eimeria praecox* is spread in exactly the same as that with *E. mitis* (see above) but is practically non-pathogenic.

Signs
Normally asymptomatic but may cause reduced feed efficiency and reduced weight gain, and predispose to other intestinal conditions.

Post-mortem lesions
Minimal, but usually excess liquid and mucus in the duodenal loop. Severe infection can cause dehydration through excessive fluid loss. The cells of the sides of the villi (not tips) are usually parasitised.

Diagnosis
Identification of characteristic slightly ovoid oocysts in the duodenum in the absence of *E. acervulina* lesions. It has a very short pre-patent period (c. 80 hours).

Treatment
Not usually treated but susceptible to the products used for other forms of intestinal coccidiosis.

Prevention
Normally controlled by anticoccidials in feed. Not usually included in vaccines.

COCCIDIOSIS, CAECAL, *E. TENELLA*

Introduction
This was at one time the commonest type of coccidiosis and is certainly the most easily diagnosed. It is caused by *Eimeria tenella* and results in lesions in the caecum of chickens worldwide. Morbidity is 10–40% and mortality up to 50%. Transmission as for *E. mitis* (see above).

Signs
Depression.
Ruffled feathers.
Closed eyes.
Inappetance.
Diarrhoea, blood in faeces.
Production less affected than in some of the other forms of coccidiosis.

Post-mortem lesions
Petechiae.
Thickening, ecchymoses, of caecal mucosa.

Accumulation of varying quantities of blood and caseous necrotic material in the caecum.

Diagnosis
Signs, lesions, microscopic examination of scrapings. Differentiate from ulcerative enteritis, histomonosis.

Treatment
Toltrazuril, Sulphonamides, Amprolium, Vitamins A and K in feed or water.

Prevention
Coccidiostats in feed, vaccination by controlled exposure, hygiene. *E. tenella* is more common when 'straight' ionophore programmes are used. Shuttle programmes with chemicals in the starter diet usually improve control. In some markets the organic arsenical compound 3-Nitro is used as an aid in the control of caecal coccidiosis. Vaccines are used mainly in breeders but increasingly in broilers. Recovered birds have good immunity to the same parasite.

Figure 15. Moderate *Eimeria tenella* infection in the caecae of a chicken (Score 3). The caecal walls are thickened and haemorrhagic and there is a mass of blood in the caecal lumen.

COCCIDIOSIS, ILEORECTAL, *E. BRUNETTI*

Introduction
A relatively rare form of coccidiosis affecting chickens worldwide caused by *Eimeria brunetti*. Of moderate to high pathogenicity, it is found in the terminal ileum, caecum and rectum. Morbidity and mortality are variable.

Signs
Depression.
Ruffled feathers.
Closed eyes.
Inappetance.
Poor production.
Diarrhoea, blood in faeces.

Post-mortem lesions
Petechiae and thickening of the distal third or more of intestine, extending into caecal tonsils.
Severe necrotising enteritis.
Oocysts in caecum and rectum.

Diagnosis
Signs, lesions, microscopic examination of scrapings. Differentiate from ulcerative enteritis, caecal coccidiosis.

Treatment
Toltrazuril, Sulphonamides, Amprolium, Vitamins A and K in feed or water.

Prevention
Coccidiostats in feed, vaccination by controlled exposure, hygiene. This species is not usually included in vaccines for broilers. There is good immunity to the same parasite in recovered birds.

Chapter 6 - Diseases and Syndromes: Chickens & Various Species

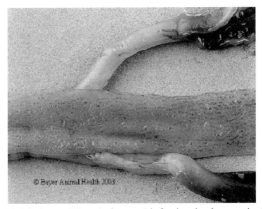

Figure 16. Moderate *Eimeria brunetti* infection in the terminal ileum and rectum of a chicken. There is thickening of the intestinal mucosa and there are lines of haemorrhagic spots in the mucosa.

COLIBACILLOSIS, COLISEPTICEMIA

Introduction

Coli-septicaemia is the commonest infectious disease of farmed poultry. It is most commonly seen following upper respiratory disease (such as Infectious Bronchitis) or Mycoplasmosis. It is frequently associated with immunosuppressive diseases such as Infectious Bursal Disease Virus (Gumboro Disease) in chickens or Haemorrhagic Enteritis in turkeys, or in young birds that are immunologically immature. It is caused by the bacterium *Escherichia coli* and is seen worldwide in chickens, turkeys, etc. Morbidity varies, mortality is 5–20%. The infectious agent is moderately resistant in the environment, but is susceptible to disinfectants and to temperatures of 80°C. Infection is by the oral or inhalation routes, and via shell membranes/yolk/navel, water, fomites, with an incubation period of 3–5 days. Poor navel healing, mucosal damage due to viral infections and immunosuppression are predisposing factors.

Signs
Respiratory signs, coughing, sneezing.
Snick.
Dejection.
Reduced appetite.
Poor growth.
Omphalitis.

Post-mortem lesions
Airsacculitis.
Pericarditis.
Perihepatitis.
Swollen liver and spleen.
Peritonitis.
Salpingitis.
Omphalitis.
Synovitis.
Arthritis.
Enteritis.
Granulomata in liver and spleen.
Cellulitis over the abdomen or in the leg.
Lesions vary from acute to chronic in the various forms of the disease.

Diagnosis
Isolation, sero-typing, pathology. Aerobic culture yields colonies of 2–5mm on both blood and McConkey agar after 18 hours – most strains are rapidly lactose-fermenting producing brick-red colonies on McConkey agar. Differentiate from acute and chronic infections with *Salmonella* spp, other enterobacteria such as *Proteus*, as well as *Pseudomonas*, *Staphylococcus* spp. etc.

Treatment
Amoxycillin, tetracyclines, neomycin (intestinal activity only), gentamycin or ceftiofur (where hatchery borne), potentiated sulphonamide, flouroquinolones.

Prevention
Good hygiene in handling of hatching eggs, hatchery hygiene,

good sanitation of house, feed and water. Well-nourished embryo and optimal incubation to maximise day-old viability. Control of predisposing factors and infections (usually by vaccination). Immunity is not well documented though both autogenous and commercial vaccines have been used.

Figure 17. Severe perihepatitis in colibacillosis in a broiler parent chicken. The liver is almost entirely covered by a substantial layer of fibrin and pus.

CONTACT DERMATITIS, HOCK BURN, PODODERMATITIS

Introduction

Contact dermatitis affects skin surfaces which have prolonged contact with litter, the foot pad, rear surface of the hock and, when severe, the breast area. It is seen in growing broiler chickens and turkeys, and in broiler parents. It is commonly monitored at processing as a means of measuring 'welfare status'. Some lesions are superficial, whereas others progress to deep ulcers so the size, stage of resolution and severity of lesion is likely to affect the degree of discomfort or pain suffered. Pododermatitis is often related to high droppings pH, level of soya bean meal in feed (according to some authors, stickiness of droppings), and, most importantly, litter moisture. It can be reproduced by adding water to the litter.

Signs

Blackened skin progressing to erosions and fibrosis on the lower surface of the foot pad, at the back of the hocks, and sometimes in the breast area. If severe the foot and hock lesions may contribute to lameness or serve as a portal of entry for secondary infections.

Post-mortem lesions

As described under signs.

Diagnosis

Signs and lesions.

Treatment

Not applicable.

Prevention

This condition is closely linked to intestinal function as this is the prime determinant of faecal consistency and stickiness. See the discussion in the section on Dysbacteriosis. Choice of drinker type (nipple as opposed to bell), drinker management, proper insulation in cold climates, and adequate ventilation to remove moisture are all important.

Figure 18. Moderate pododermatitis on a foot pad.

CROPWORMS

Introduction

The nematode worms *Capillaria* spp and *Gongylonema ingluvicola* infect the mucosa of the crop and oesophagous of poultry and game birds. Some have beetle or earthworms as intermediate hosts.

Signs
Anaemia.
Emaciation.

Post-mortem lesions
Inflammation and thickening of mucosa of crop and oesophagus. White convoluted tracks in the mucosa.

Diagnosis
Microscopic examination of mucosal scraping.

Treatment
Levamisole, Coumaphos.

Prevention
Effective cleaning of housing. Avoidance of access to intermediate hosts. Routine worming.

CRYPTOSPORIDIOSIS

Introduction
Cryptosporidia are related to the coccidia, but much smaller (typically oocysts are less than ¼ of the size of an *E. acervulina* oocyst). They replicate in the brush border on the surface of epithelial cells. They also differ from coccidia in being poorly host specific, although bird strains do not infect mammals very well, and vice versa. *Cryptosporidium baileyi* can cause respiratory disease in chickens and turkeys. The same species causes infections of the hindgut and cloacal bursa in chickens, turkeys, and ducks. *C. meleagridis* also infects both species. A further species causes respiratory disease in quail. The oocysts are excreted ready sporulated in the faeces and infection occurs by inhalation and ingestion.

Signs
Snick.
Cough.
Swollen sinuses.

Low weight gain.
Diarrhoea.

Post-mortem lesions
Sinusitis.
Airsacculitis.
Pneumonia.

Diagnosis
Identification of the parasites attached to the epithelium by microscopic examination (smears histopathology acid-fast staining)

Treatment
Unfortunately there is currently no known effective treatment in poultry. If other disease processes are complicating the situation (e.g. coli-septicaemia) there may be benefit in medicating for these.

Prevention
The oocysts of cryptosporidia are extremely resistant to chemical disinfection. There are no effective preventative medicines or feed additives. It is becoming increasingly common for water companies to screen water supplies for cryptosporidia because of the human health implications of mammalian strains. Steam cleaning is effective in reducing infection as oocysts are inactivated above about 65°C.

DACTYLARIOSIS

Introduction
A rare fungal disease of chickens and turkeys caused by *Dactylaria gallopava*.

Signs
Incoordination.
Tremors.
Torticollis.
Circling, recumbency.

Post-mortem lesions
Necrotic lesions with associated congestion in cerebrum.

Mycotic lesions in lungs, air sacs etc.

Diagnosis
Lesions, isolation of the fungus.

Treatment
None.

Prevention
Use fresh dry litter (avoid old sawdust).

DEGENERATIVE JOINT DISEASE

Introduction
A condition of chickens and turkeys seen worldwide. The cause remains to be confirmed, but it may result from physical damage, or developmental defects. Morbidity and mortality are low but affected birds are more likely to be 'picked upon' and may end up suffering damage and needing to be culled. Rapid growth is a possible predisposing factor.

Signs
Lameness.
Reduced breeding performance.

Post-mortem lesions
Damaged epiphyseal articular cartilage, especially of femoral anti-trochanter but also other leg joints, resulting in erosions, and cartilage flaps.
Microscopically there is necrosis and there may be fissures of articular cartilage and associated osteochondritis.

Diagnosis
Gross and microscopic lesions.

Treatment
None available. Appropriate management of a segregation pen and early marketing of mildly affected birds may limit losses and improve flock welfare.

Prevention
Avoidance of physical sources of injury to bones and joints.

There may be a place for growth-control programmes, especially during period of rapid growth.

DEPLUMING AND SCALY LEG MITES

Introduction
External parasites of adult chickens, pheasants, pigeons etc, *Knemidocoptes* spp.

Signs
Cause irritation and the bird pulls feathers.
Mange lesions on legs and unfeathered parts.
Unthriftiness.
Raised thickened scales.

Post-mortem lesions
As described under signs.

Diagnosis
Signs, microscopic examination for mites in scrapings. The adult females are short legged, round, up to 0.5mm in diameter.

Treatment
Not usually required in commercial poultry. For small flocks dipping the affected parts in a solution of acaricide may be beneficial. Application of mineral or vegetable oil is also beneficial.

Prevention
Careful cleaning of buildings during down-time will help reduce the risk of these infections. Exclusion of wild birds from chicken areas is advised as far as possible. It may be best to cull affected birds from small flocks.

DYSBACTERIOSIS, NON-SPECIFIC BACTERIAL ENTERITIS

Introduction
Inflammation of the small intestine associated with wet litter, excess

caecal volume and fermentation, is common in countries with restrictions on the use of antimicrobial growth promoters and pressure to reduce therapeutic antimicrobial usage. The condition is seen mainly in rapidly growing broiler chickens with good food intake. Dietary changes, feed interruptions and subclinical coccidiosis may be contributory factors. No single bacterium appears to be responsible, rather we are dealing with a disruption in the normal flora of the gut.

Signs
Diarrhoea.
Water intake may be increased or irregular.

Post-mortem lesions
Excessive fluid content throughout the small intestine.
Wet faeces in the rectum.
Voluminous caecae, often with gas bubbles.

Diagnosis
Signs, lesions, microscopic examination of scrapings from the wall of the small intestine (to exclude coccidiosis and perhaps to assess the bacterial flora).

Treatment
Amoxycillin and tylosin treatment appear to be beneficial, especially where treatment is initiated early. Treatment should coincide with good relittering and it is important to provide fresh sanitary drinking water.

Prevention
Competitive exclusion ('normal adult flora') use in day-old chicks reduces the risk of this condition. Feed acidification may be helpful in some circumstances. Careful choice of any feed enzymes and their matching with local raw materials can have an impact on substrates made available to intestinal bacteria. Good control of coccidiosis. Prophylactic antimicrobial medication may be necessary in some circumstances.

Egg drop syndrome 76

Introduction

Egg drop may be defined as a sudden drop in egg production or a failure to achieve a normal peak in production. In the autumn of 1976 a distinct egg drop syndrome was first identified in Northern Ireland. Apparently a similar disease had been seen over a 4-year period in broiler parents in Holland. The cause has been identified as Adenovirus BC14, 127, first isolated in Northern Ireland in 1976. It affects chickens and has occurred in Ireland, Holland, France, England, Germany, Spain, Peru, Brazil, Uruguay, Argentina. Mortality is usually negligible. Circumstantial evidence suggests that the main route of transmission is through the eggs (vertical transmission) followed by latent infection during rear with viral excretion starting shortly before sexual maturity. Lateral transmission from bird to bird is slow and may be prevented or slowed for weeks by netting divisions. Contamination of egg trays at packing stations may play a part in transmission, as may wildfowl and biting insects. Clinical disease occurs during sexual maturity. Spread from house to house may take 5–10 weeks. Unvaccinated flocks with antibodies before lay do not peak normally. The infection is commonly present in ducks and geese but does not cause disease.

Signs

Egg drop at peak or failure to peak. Drops may be of 5 to 50% and last for 3–4 weeks.

Rough, thin or soft-shelled eggs and shell-less eggs.

Loss of shell pigment.

Poor internal quality.

Lack of signs in the birds themselves.

Post-mortem lesions

No specific lesion – only a slight atrophy of ovary and oviduct.

Histopathology – it may be possible to demonstrate degenerative changes in the epithelial cells of the magnum of the oviduct.

Diagnosis

History, signs/lesions (mainly lack of). Isolation of

haemagglutinatin agent in duck eggs or cell culture, group antigen distinct from classical adenoviruses (white cells, throat swabs, oviduct). Serology: HI, SN, DID, Elisa. It is important to rule out other possible reasons for egg drop, which can be caused by a large number of factors acting individually or in combination. Management problems may be involved: inadequate water supply; extremes of temperature; inadequate lighting programme; sudden changes of feed. Nutritional deficiency should be considered, specifically vitamins E, B_{12} and D as well as calcium, phosporus, selenium. Diseases in which egg drop occurs, may be infectious or metabolic. Infectious diseases include Infectious Bronchitis, Infectious Laryngotracheitis, Avian Encephalomyelitis, Newcastle disease, Marek's disease/Leukosis or any infectious disease causing a significant systemic disturbance (CRD, Coryza, Cholera, Parasites, Diphtheritic Fowl Pox). Metabolic diseases include Fatty Liver Syndrome, intoxication by sulphonamides, insecticides or nicarbazin.

Treatment
None. Soluble multivitamins may be recommended as a non-specific measure.

Prevention
Vaccination with inactivated vaccine prior to lay.

ENDOCARDITIS

Introduction
A condition of chickens associated with several bacterial infections such as staphylococci, streptococci, *Erysipelothrix* etc.

Signs
Fluid-distended abdomen.
Peripheral vessels congested.

Post-mortem lesions
Right ventricular failure and ascites.
Vegetative lesions usually on the right atrioventricular valve.

Diagnosis
Differentiate from broiler ascites syndrome by examination of the interior of the heart. Culture of lesions to confirm the bacterium involved.

Treatment
Medication is only likely to be of value in reducing deterioration in birds that are starting to develop lesions. The choice should depend on sensitivity testing of an isolate.

Prevention
Good hygiene at turn-around. If successive flocks are affected on the same site prophylactic medication ahead of the anticipated problem may be of benefit.

EPIPHYSIOLYSIS

Introduction
A complex condition of chickens that may be associated with trauma, growth plate disease, rickets, bacterial infection, osteomyelitis, and /or trauma.

Signs
Apparent dislocation at extremities of long bones.

Post-mortem lesions
Separation of epiphyses at the growth plate, sometimes seen *in vivo*, often occurs or is identified in the processed carcase.

Diagnosis
Gross inspection, histology may be helpful in identifying underlying problems such as rickets or 'FHN'.

Treatment
Not applicable.

Prevention
Control of predisposing factors.

Equine encephalomyelitis (EEE, WEE, VEE, eastern, western or Venezuelan)

Introduction
A viral disease of pheasants, partridges, wild birds, chickens, turkeys, ducks and pigeons having a high morbidity and high mortality. It is transmitted between birds by pecking and by mosquitoes. These conditions currently only occur from northern South America to North America. The natural hosts are wild birds and rodents. Horses are also seriously affected.

Signs
Nervous symptoms.
Ataxia.
Paresis.
Paralysis.
Flaccid neck.
Circling.
Tremors.
May also be asymptomatic.

Post-mortem lesions
No gross lesions.
Microscopic lesions not pathognomonic.

Diagnosis
Isolation in mice, TC, and CE, ID by VN, COFAL.

Treatment
None.

Prevention
Protection from mosquitoes, control cannibalism. Vaccinate at 5–6 week.

Erysipelas

Introduction
A sudden onset infection with the bacterium *Erysipelothrix*

insidiosa (*E. rhusiopathiae*) seen in turkeys and increasingly in free-range chickens, rarely in geese, ducks, pheasants. It is also seen in some mammals. It may be transmitted by faecal carriers for 41 days, in soil, water, fishmeal and semen and by cannibalism. The bacterium is fairly resistant to environmental effects or disinfectants and may persist in alkaline soil for years. There is likely to be an increased risk if housing or land has been previously used by pigs or sheep.

Signs

Inappetance.
Depression.
Sleepiness.
Swollen snood.
May be diarrhoea and respiratory signs.
Perineal congestion.
Chronic scabby skin, especially snood.
Sudden death.

Post-mortem lesions

Carcase congestion.
Liver, kidney, spleen swollen.
Haemorrhages in fat, muscle, epicardium.
Marked catarrhal enteritis.
Joint lesions.
Endocarditis.

Diagnosis

Isolation on blood agar, and identification; the demonstration of the organism in stained impression smears from tissues. Vaccination or natural infection may cause false positive reactions in the *Mycoplasma gallisepticum* and *M. synoviae* plate tests for a few weeks. Differentiate from pasteurellosis, salmonellosis, colibacillosis, and acute Newcastle disease.

Treatment

Penicillin – a combination of the procaine and benzathine salts may be injected, often along with bacterin. Tetracyclines in feed may also be helpful.

Prevention
Good biosecurity to prevent spread from other susceptible species, vaccine at 16–20 weeks if the condition is enzootic.

FATTY LIVER HAEMORRHAGIC SYNDROME

Introduction
A condition occurring worldwide in chickens, especially caged layers and with a complex set of causes including excessive calories, mycotoxins, deficiency and stress.

Signs
Overweight typically by 25%.
Sudden death.
Sudden drop in egg production.
Some birds with pale comb and wattles.

Post-mortem lesions
Obesity.
Headparts pale.
Liver yellow, greasy and soft with numerous haemorrhages.
Death by internal exsanguination after rupture of haematocyst.

Diagnosis
Lesions, history.

Treatment
Reduce energy intake, supplement with choline, vitamin E, B_{12} and inositol.

Prevention
Feed to avoid obesity, avoid mycotoxins and stress.

FAVUS

Introduction
A fungal infection, *Trichophyton gallinae*, of chickens and turkeys. It is very rare in commercial poultry production.

Signs
White, powdery spots and wrinkled crusts and scab on comb and wattles.
Feather loss.
'Honeycomb' skin.
Thick crusty skin.
Loss of condition.

Post-mortem lesions
See signs (above).

Diagnosis
Lesions, isolation.

Treatment
Formalin in petroleum jelly.

Prevention
Good hygiene of facilities, culling affected birds.

'Femoral Head Necrosis', FHN, Bacterial chondronecrosis with osteomyelitis, 'Proximal Femoral Degeneration'

Introduction
A condition of chickens and turkeys that may be associated with several different bacterial infections e.g. staphylococci, *E. coli*, streptococci. FHN is the commonest infectious cause of lameness in broilers in the UK. Post-mortem studies of birds culled due to lameness and of birds found dead, indicated that 0.75% of all male broilers placed had lesions in the hip bone. Predisposing factors include immunosuppresive viruses such as Infectious Bursal Disease Virus and Chicken Anaemia Virus and non-infectious bone pathologies such as hypophosphaetamic rickets.

Signs
Lameness.
Use of a wing for support during walking and hip flexion.

Post-mortem lesions
Degeneration of the epiphyses of long bones with thinning of the cortex and tendency to break when force is applied.

Diagnosis
Base on post-mortem lesions and isolation of a causative organism. Differentiate from synovitis, arthritis, spondylolisthesis.

Treatment
Antibiotic therapy in accordance with sensitivity is likely to be beneficial only for birds in the early stage of this process and may not be economically justifiable.

Prevention
Exclusion of floor eggs and dirty eggs from the hatchery. Careful attention to mineral and Vitamin D nutrition to avoid subclinical, especially hypophosphataemic, rickets.

FOWL CHOLERA, PASTEURELLOSIS

Introduction
Fowl Cholera is a serious, highly contagious disease caused by the bacterium *Pasteurella multocida* in a range of avian species including chickens, turkeys, and water fowl, (increasing order of susceptibility). It is seen worldwide and was one of the first infectious diseases to be recognised, by Louis Pasteur in 1880. The disease can range from acute septicaemia to chronic and localised infections and the morbidity and mortality may be up to 100%. The route of infection is oral or nasal with transmission via nasal exudate, faeces, contaminated soil, equipment, and people. The incubation period is usually 5–8 days. The bacterium is easily destroyed by environmental factors and disinfectants, but may persist for prolonged periods in soil. Reservoirs of infection may be present in other species such as rodents, cats, and possibly pigs. Predisposing factors include high density and concurrent infections such as respiratory viruses.

Signs
Dejection.
Ruffled feathers.
Loss of appetite.
Diarrhoea.
Coughing.
Nasal, ocular and oral discharge.
Swollen and cyanotic wattles and face.
Sudden death.
Swollen joints.
Lameness.

Post-mortem lesions
Sometimes none, or limited to haemorrhages at few sites.
Enteritis.
Yolk peritonitis.
Focal hepatitis.
Purulent pneumonia (especially turkeys).
Cellulitis of face and wattles.
Purulent arthritis.
Lungs with a consolidated pink 'cooked' appearance in turkeys.

Diagnosis
Impression smears, isolation (aerobic culture on trypticase soy or blood agar yields colonies up to 3mm in 24 hours – no growth on McConkey), confirmed with biochemical tests. Differentiate from Erysipelas, septicaemic viral and other bacterial diseases.

Treatment
Sulphonamides, tetracyclines, erythromycin, streptomycin, penicillin. The disease often recurs after medication is stopped, necessitating long-term or periodic medication.

Prevention
Biosecurity, good rodent control, hygiene, bacterins at 8 and 12 weeks, live oral vaccine at 6 weeks.

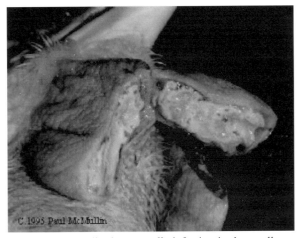

Figure 19. Severe localised *Pasteurella* infection in the swollen wattle of a 30-week-old male broiler parent chicken. The swelling is made up of oedema and purulent exudates (pus).

FOWL PLAGUE, AVIAN INFLUENZA-HIGHLY PATHOGENIC (HPAI)

Introduction

One of only two 'Class A' diseases of poultry targeted for emergency disease control measures by OIE, the equivalent of the World Health Organisation for animal diseases. This viral disease can cause exceptionally high mortality, especially in turkeys. In addition official control measures disrupt trade in poultry products from affected areas. The cause is a virus, Orthomyxovirus type A, its pathogenicity is variable, and isolates are designated sero-type/species/location/reference number/year/subtype designation(H/N). Highly pathogenic forms are usually of the H groups 5 and 7 and may now be identified (if H5 or H7) by the presence of a sequence at the haemagglutinin cleavage site that codes for multiple basic amino acids. The definitive classification of high pathogenicity is an intravenous pathogenicity test (IVPI) in 6-week-old chickens result of greater than 1.2 . This is a test in which the virus is inoculated into susceptible chickens that are then kept under observation. The higher the proportion of the chickens dying or showing signs the higher the IVPI. The virus infects chickens, turkeys, ducks, partridges, pheasants, quail, pigeons, and ostriches. Effectively all birds are considered to be at risk of infection. Apathogenic and mildly pathogenic influenza A viruses occur worldwide. Highly pathogenic avian influenza A (HPAI) viruses of the H5 and H7 HA subtypes have been isolated occasionally from free-living birds. Outbreaks due to HPAI were recorded in the Pennsylvania area, USA, in the years 1983–84. More recently outbreaks have occurred in Australia, Pakistan, Mexico and, from December 1999, in northern Italy. A serious outbreak occurred in The Netherlands in 2003 with a few linked cases in Belgium and one in Germany. H5 viruses of low pathogenicity may become highly pathogenic usually after circulating in poultry flocks for a time (Pennsylvania, Italy). Because of this, and the high mortality that 'low-path' AI can cause in turkeys, OIE and other bodies are currently examining ways to improve control of LPAI. See current OIE records for up to date information on distribution of HPAI. Morbidity is high but mortality usually relatively low, 5–50%. The route of infection is probably oral initially, but possibly by the conjunctival or respiratory route and the incubation period is 3–5 days. Transmission is by direct contact with secretions from infected birds, especially faeces, waterfowl, equipment, clothing, drinking water. The virus replicates mainly in respiratory tissues of chickens and turkeys but in the intestinal tract of clinically normal waterfowl. Avirulent in one species may be virulent in others. Broken contaminated eggs may infect chicks in the incubator simulating vertical transmission. The virus is moderately resistant, can survive 4 days in water at 22°C, over 30 days at 0°C. It is inactivated by a temperature of 56°C in 3 hours and 60°C in 30 min, by acid pH, by oxidising agent and by formalin and iodine compounds. It can remain viable for long periods in tissues. Infections with other pathogens (e.g. *Pasteurella*) may increase mortality, even with 'low pathogenicity' strains.

Avian Influenza is a potential zoonosis. It can result in inapparent infection, conjunctivitis or severe pneumonia. The small number of human deaths associated with HPAI appear to have resulted from direct exposure to infected birds on farm or in markets.

Signs
Sudden death.
Marked loss of appetite, reduced feed consumption.
Cessation of normal flock vocalisation.
Drops in egg production.
Depression.
Coughing.
Nasal and ocular discharge.
Swollen face.
Cyanosis of comb/wattles.
Diarrhoea (often green).
Nervous signs such as paralysis.

Post-mortem lesions
Inflammation of sinuses, trachea, air sacs and conjunctiva.
Ovarian regression or haemorrhage.
Necrosis of skin of comb and wattles.
Subcutaneous oedema of head and neck.
Dehydration.
Muscles congested.
Haemorrhage in proventricular and gizzard mucosae and lymphoid tissue of intestinal tract.
Turkey lesions tend to be less marked than those of chickens, while ducks may be symptomless, lesionless carriers of highly pathogenic virus.

Diagnosis
A presumptive diagnosis may be made on history and post-mortem lesions. Confirmation is by viral isolation in chick embryo, HA+, NDV-, DID+. Commercial Elisa test kits are now available. However, as with many such tests occasional false positive reactions can occur. The agar gel precipitation test is non-group-specific and is used to confirm any positives. Differentiate from Newcastle disease, fowl cholera, infectious laryngotracheitis, other respiratory infections, bacterial sinusitis in ducks.

Treatment

None, but good husbandry, nutrition and antibiotics may reduce losses. Eradication by slaughter is usual in chickens and turkeys.

Prevention

Hygiene, quarantine, all-in/all-out production, etc. Minimise contact with wild birds, controlled marketing of recovered birds. Vaccination is not normally recommended because, although it may reduce losses initially, vaccinated birds may remain carriers if exposed to the infection. Vaccines have been used in recent outbreaks in Mexico and Pakistan. To be effective inactivated vaccines must be the right subtype for the particular situation (H5 will not protect against H7 and vice versa). In outbreaks a regime of slaughter, correct disposal of carcases, cleaning, disinfection, isolation, 21-day interval to re-stocking should be followed. Survivors can be expected to have a high degree of immunity but may harbour virulent virus.

FOWL POX, POX, AVIAN POX

Introduction

A relatively slow-spreading viral disease characterised by skin lesions and/or plaques in the pharynx and affecting chickens, turkeys, pigeons and canaries worldwide. Morbidity is 10–95% and mortality usually low to moderate, 0–50%. Infection occurs through skin abrasions and bites, or by the respiratory route. It is transmitted by birds, fomites, and mosquitoes (infected for 6 weeks). The virus persists in the environment for months. It is more common in males because of their tendency to fight and cause skin damage, and where there are biting insects. The duration of the disease is about 14 days on an individual bird basis.

Signs

Warty, spreading eruptions and scabs on comb and wattles.
Caseous deposits in mouth, throat and sometimes trachea.
Depression.
Inappetance.
Poor growth.
Poor egg production.

Post-mortem lesions

Papules progressing to vesicles then pustules and scabs with distribution described above.

Less commonly there may, in the diptheritic form, be caseous plaques in mouth, pharynx, trachea and/or nasal cavities.

Microscopically – intra-cytoplasmic inclusions (Bollinger bodies) with elementary bodies (Borrel bodies).

Diagnosis

A presumptive diagnosis may be made on history, signs and post-mortem lesions. It is confirmed by IC inclusions in sections/scrapings, reproduction in susceptible birds, isolation (pocks on CE CAM) with IC inclusions. DNA probes. Differentiate from Trichomoniasis or physical damage to skin.

Treatment

None. Flocks and individuals still unaffected may be vaccinated, usually with chicken strain by wing web puncture. If there is evidence of secondary bacterial infection broad-spectrum antibiotics may be of some benefit.

Prevention

Vaccination (except canary). Chickens well before production. Turkeys by thigh-stick at 2–3 months, check take at 7–10 days post vaccination. There is good cross-immunity among the different viral strains.

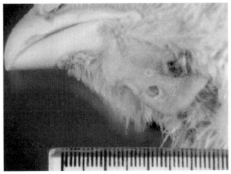

Figure 20. Fowl pox lesions on the wattle of an adult broiler parent chicken.

Gangrenous Dermatitis, Necrotic dermatitis

Introduction
A bacterial condition seen in chickens and usually caused by *Clostridium septicum*, occasionally *Staphylococcus aureus*, rarely *Clostridium noyvi/oedematiens*. Morbidity may be up to 50% and mortality is high. It occurs due to invasion of 'normal' wounds by 'normal' bacteria in immunosuppressed birds. Immunosuppression is therefore a predisposing factor, especially following congenital Chick Anaemia Virus infection or early Infectious Bursal Disease Virus infection (Gumboro Disease). The spores of Clostridial bacteria are highly resistant in the environment.

Signs
Occasionally dejection.
Loss of appetite.
Gangrenous skin.
Severe cellulitis especially of thighs, wings, wattles.
Sudden mortality.

Post-mortem lesions
Patches of gangrenous skin with underlying emphysematous and/or sanguinous cellulitis, usually over wings and breast, sometimes thighs and other parts.
Swelling and infarction of the liver, spleen.
Foci in liver.

Diagnosis
Clinical signs and/or lesions. Recovery of an abundant growth of causative organism in recently dead birds.

Treatment
Sulphaquinoxaline, penicillin or amoxycillin.

Prevention
Good hygiene and management. Avoid skin trauma and immunosuppression (congenital CAV infection, early Infectious Bursal Disease Virus infection).

Figure 21. Severe gangrenous dermatitis on the upper surface of the wing of a broiler chicken.

GAPE

Introduction
Syngamus trachea, a nematode worm parasite of chickens, turkeys, pheasants, and other game and ornamental birds occurring worldwide. Infection is by the oral route with earthworms, slugs and snails acting as transfer hosts but the life cycle may also be direct, by ingestion of embryonated egg or L3. There is an 18–20 day prepatent period. The condition is seen more commonly in poultry on free range where ground may be contaminated by wild birds e.g. from rookeries.

Signs
    Gaping.
    Dyspnoea.
    Head shaking.
    Loss of appetite and condition.

Post-mortem lesions
    Tracheitis.
    Presence of worms, paired parasites up to 2 cm long.

Diagnosis
    Signs and lesions, confirmation of presence of the parasite.

Treatment
Flubendazole in feed, levamisole.

Prevention
Flubendazole.

GIZZARD WORMS – CHICKENS

Introduction
Cheilospirura, Streptocara, and *Histiocephalus* are nematode worm parasites of chickens. *Cheilospirura* and *Streptocara* are seen worldwide but *Histiocephalus* is restricted to Europe. Grasshoppers, weevils, beetles etc act as intermediate hosts. They are more common in free-range birds because of their increased access to intermediate hosts.

Signs
Depression.
Loss in condition and weight.
Slow growth.

Post-mortem lesions
Ulceration, necrosis and partial sloughing of gizzard lining, muscular wall may be sacculated or ruptured.
Adults are 2–4 cm long and usually bright red.

Diagnosis
Lesions, confirmation of presence of the worms.

Treatment
Levamisole, benzimidazoles such as flubendazole.

Prevention
Prevention of access to intermediate hosts, routine worming.

HEAT STRESS

Introduction
A condition seen in chickens, and turkeys caused by high environmental temperature, especially associated with high relative

humidity and low air speed. Ducks are relatively resistant to heat stress. Predisposing factors include genetics, feather cover, high stocking density, nicarbazin in feed, acclimation, drinking water temperature and availability.

Signs
Panting.
Increased thirst.
Reduced feed consumption.
Reduced egg production.
Legs and wings outstretched.
Prostration.

Post-mortem lesions
Carcases congested.
Mucoid exudate in nostrils and mouth.

Diagnosis
Temperature records, signs, lesions, exclusion of other conditions, pattern of losses.

Treatment
Cool water, maximise airflow, if relative humidity is low then wet the roof and fog.

Prevention
Houses of optimal height and insulation, painted white to reflect heat, evaporative coolers, feed with a reduced protein:energy ratio. Feeding during cooler hours may be beneficial. Reductions in stocking density of meat poultry may be quickly achieved by partial depletion ('thinning').

HAEMORRHAGIC DISEASE, APLASTIC ANAEMIA, HAEMORRHAGIC ANAEMIA

Introduction
A complex condition of chickens associated with drug toxicities, mycotoxins and viral infections and usually following a course of approximately 3 weeks. Morbidity varies, mortality is 5–50%.

Signs
Dejection.
Loss of appetite.
Poor growth.
Pale comb and wattles.
Blood in droppings.

Post-mortem lesions
Haemorrhages in one or more sites: skin, muscles, liver, heart, serosa and mucosae.
Liver yellow.
Carcase anaemia.
Bone marrow pale with fatty change.

Diagnosis
History, signs, lesions.

Treatment
Vitamin K, remove sulphonamides, add liver solubles to feed.

Prevention
Avoid causative factors.

HYDROPERICARDIUM-HEPATITIS SYNDROME, ANGARA DISEASE

Introduction
This condition was first identified in broilers in Pakistan in 1987. It spread rapidly in broiler producing areas in that country and the same or a very similar condition has been seen in North and South America. It affects mainly broilers and broiler parents in rear and has also been seen in pigeons. It is a condition caused by an adenovirus, possibly in combination with an RNA virus and immunosuppression caused by Chick Anaemia Virus or Infectious Bursal Disease. The disease is readily reproduced by inoculating birds with a bacteria-free filtrate of a liver extract from an affected bird. Mortality may reach 60% but more typically 10–30%.

Signs
Sudden increase in mortality.
Lethargy.
Huddling with ruffled feathers.
Yellow mucoid droppings.

Post-mortem lesions
Excessive straw-coloured fluid distending the pericardium (up to 10 mls).
Enlarged, pale friable liver and kidney.
Congestion of the carcase.
Lungs oedematous.

Diagnosis
Lesions, histopathology, virology.

Treatment
None. Good water sanitation (e.g. treatment of drinking water with 0.1% of a 2.5% iodophor solution) appears to be beneficial.

Prevention
The condition typically occurs in areas of high poultry density where multi-age operation is traditional. Control of predisposing immunosuppressive diseases may help limit losses. Formalin-inactivated oil adjuvant vaccines are reported to be highly effective and are used in areas where the condition is endemic.

IMPACTION AND FOREIGN BODIES OF GIZZARD

Introduction
The gizzard or crop may become impacted with litter, grass, string etc. The normal function of the gizzard is to aid in the physical grinding of food materials, to reduce their particle size to aid digestion. Gizzard activity also acts as a pacemaker of intestinal activity and controls the speed at which food is passed to the small intestine. Most young commercial poultry consume feeds that have a small particle size. Older birds ingest grit to facilitate the grinding activity in the gizzard. This condition usually affects only a small number of birds, however if young chicks to do not begin to eat

feed properly they often consume litter instead. Impacted gizzards are then found in 'non-starter' type chicks or poults. Grit would not be classed as a foreign body, however sometimes free-range poultry consume large stones, and birds of any age can consume nails, staples etc. This usually happens after maintenance activities have been carred out in the housing.

Signs
Reduced feed intake.
Reduced weight gain.

Post-mortem lesions
The gizzard is more firm than normal and, and on opening is found to contain a mass of fibrous material. This may extend into the proventriculus and on into the duodeunum.
A foreign body may be found in the interior of the gizzard.
Nails commonly penetrate the lining of the gizzard, and may penetrate the body wall.

Diagnosis
Lesions.

Treatment
None effective in young birds. Obvious non-starters should be culled in this situation.

Prevention
Good brooding management to encourage early adaptation to the diet and adequate consumption of feed and water. Daily monitoring of the crop-fill is a useful procedure. It allows us to confirm the proportion of chicks consuming feed and can give inexperienced poultry keepers feedback on the effectiveness of their brooding management. Sweep floors after maintenance activities to remove nails and other potentially dangerous foreign bodies.

INCLUSION BODY HEPATITIS

Introduction
A disease of chickens characterised by acute mortality, often with

severe anaemia, caused by an adenovirus. A number of different sero-types have been isolated from disease outbreaks but they may also be isolated from healthy chickens. The disease was first described in the USA in 1963 and has also been reported in Canada, the UK, Australia, Italy, France and Ireland. It has a course of 9–15 days with a morbidity of 1–10% and a mortality of 1–10%. Infected birds remain carriers for a few weeks. Transmission may be vertical or lateral and may involve fomites. Immunosuppression, for instance due to early IBD challenge or congenital CAV infection, may be important. The virus is generally resistant to disinfectants (ether, chloroform, pH), and high temperatures. Formaldehyde and iodides work better. Curiously, many different sero-types have been isolated from different cases, but on the other hand many field cases show eosinophilic inclusions that do not appear to have adenovirus particles. Since adenoviruses are commonly found in healthy poultry, isolation alone does not confirm that they are the cause of a particular problem. A form of the disease affecting birds under 3 weeks of age in Australia has been reproduced with sero-type 8 adenovirus. Progeny of high health status breeding flocks appear to be at greater risk, perhaps because they have lower levels of maternal antibody.

Signs

Depression.
Inappetance.
Ruffled feathers.
Pallor of comb and wattles.

Post-mortem lesions

Liver swollen, yellow, mottled with petechiae and ecchymoses.
Kidneys and bone marrow pale.
Blood thin.
Bursa and spleen small.
Microscopically – basophilic intranuclear inclusions.

Diagnosis

A presumptive diagnosis may be made on history and lesions. Confirmation is made on finding inclusions in the liver. The virus grows well in tissue culture (CEK, CEL) Serology: DID for group antigen, SN for individual sero-types. Differentiate from Chick anaemia syndrome, sulphonamide intoxication, Infectious Bursal Disease, vibrionic hepatitis, fatty liver syndrome, and deficiency of vitamin B_{12}.

Treatment
None. Soluble multivitamins may help with the recovery process.

Prevention
Quarantine and good sanitary precautions, prevention of immunosuppression.

INFECTIOUS BRONCHITIS, IB

Introduction
This infection, probably the commonest respiratory disease of chickens, was first described in the USA (N. Dakota, 1931). Its affects vary with: the virulence of the virus; the age of the bird; prior vaccination; maternal immunity (young birds); and complicating infections (*Mycoplasma*, *E. coli*, Newcastle disease). Morbidity may vary 50–100% and mortality 0–25%, depending on secondary infections. The cause is a Coronavirus that is antigenically highly variable; new sero-types continue to emerge. About eight sero-groups are recognised by sero-neutralization. Typing by haemagglutination-inhibition is also used. These differences are due to structural differences in the spike proteins (S1 fraction).

Infection is via the conjunctiva or upper respiratory tract with an incubation period of 18–36 hours. The infection is highly contagious and spreads rapidly by contact, fomites or aerosol. Some birds/viral strains can be carriers to 1 year. The virus, which may survive 4 weeks in premises, is sensitive to solvents, heat (56°C for 15 mins), alkalis, disinfectants (Formal 1% for 3 mins). Poor ventilation and high density are predisposing factors.

Signs
Depression.
Huddling.

Loss of appetite.
Coughing, gasping, dyspnoea.
Wet litter.
Diarrhoea.
Diuresis.

Post-mortem lesions
Mild to moderate respiratory tract inflammation.
Tracheal oedema.
Tracheitis.
Airsacculitis.
Caseous plugs in bronchi.
Kidneys and bronchi may be swollen and they and the ureters may have urates.

Diagnosis
Tentative diagnosis is based on clinical sgns, lesions and serology. Definitive diagnosis is based on viral isolation after 3–5 passages in chick embryo, HA negative, with typical lesions, flourescent antibody positive and ciliostasis in tracheal organ culture. Serology: HI, Elisa (both group specific), SN (type specific), DID (poor sensitivity, short duration, group specific). Differentiate from Newcastle disease (lentogenic and mesogenic forms), mycoplasmosis, vaccinal reactions, Avian Influenza and Laryngotracheitis.

Treatment
Sodium salicylate 1gm/litre (acute phase) where permitted – antibiotics to control secondary colibacillosis (q.v.).

Prevention
Live vaccines of appropriate sero-type and attenuation, possible reactions depending on virulence and particle size. Maternal immunity provides protection for 2–3 weeks. Humoral immunity appears 10–14 days post vaccination. Local immunity is first line of defence. Cell-mediated immunity may also be important.

Infectious Bronchitis, IB – 793B variant
Sudden Death Syndrome in broiler parents

Introduction
A Coronavirus infection of chickens with a morbidity of 50–100% and a mortality 0–25%, depending on secondary infections. Infection is via the conjunctiva or upper respiratory tract with an incubation period of 18–36 hours. The infection spreads rapidly by contact, fomites or aerosol. Some birds/viral strains can be carriers for up to 1 year. The virus, which may survive 4 weeks in premises, is sensitive to solvents, heat (56°C for 15 mins), alkalis, disinfectants (Formal 1% for 3 mins). Poor ventilation and high density are predisposing factors.

Signs
Sudden death.
Muscular shivering.
Otherwise as for standard IB.

Post-mortem lesions
Oedema of pectoral muscles and subcutaneously on abdomen, lesions progress to necrosis and scarring of deep pectorals in convalescence.

In layers the ovules may be intensely congested.

Other lesions of 'classical' IB may be encountered.

Diagnosis
3–5 passages in CE allantoic cavity, HA-, typical lesions, FA, ciliostatic in tracheal organ culture, cell culture (Vero, CK) only after adaptation Serology: HI, Elisa (both group specific), SN (type specific), DID (poor sensitivity, short duration, group specific).

Treatment
Sodium salicylate 1gm/litre (acute phase) where permitted – antibiotics to control secondary colibacillosis (q.v.).

Prevention
Live vaccines of appropriate sero-type and attenuation, possible reactions depending on virulence and particle size.

INFECTIOUS BRONCHITIS, IB EGG-LAYERS

Introduction
A Coronavirus infection of chickens, with much antigenic variation. The condition has a morbidity of 10–100% and mortality of 0–1%. Infection is via the conjunctiva or upper respiratory tract. There is rapid spread by contact, fomites or aerosol. A few birds are carriers up to 49 days post infection. The virus is moderately resistant and may survive 4 weeks in premises. Poor ventilation and high density are predisposing factors.

Signs
Drop in egg production (20–50%).
Soft-shelled eggs.
Rough shells.
Loss of internal egg quality.
Coughing, sneezing.
Rales may or may not be present.

Post-mortem lesions
Follicles flaccid.
Yolk in peritoneal cavity (non-specific).

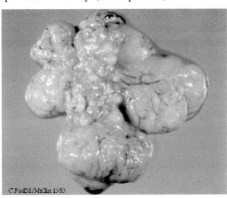

Figure 22. Flaccid ovarian follicles in a broiler parent chicken undergoing challenge with Infectious Bronchitis virus.

Diagnosis
3–5 passages in CE, HA-, typical lesions, FA. Serology: HI, SN, Elisa, DID. Differentiate from Egg Drop Syndrome, EDS-76.

Treatment
Sodium salicylate 1gm/litre (acute phase) where permitted – antibiotics to control secondary colibacillosis (q.v.).

Prevention
Live vaccines of appropriate sero-type and attenuation, although reactions can occur depending on prior immunity, virulence, particle size (if sprayed) and general health status. Maternal immunity provides protection for 2–3 weeks. Humoral immunity appears 10–14 days post vaccination. Local immunity is the first line of defence. Cell-mediated immunity may also be important.

INFECTIOUS BURSAL DISEASE, IBD, GUMBORO

Introduction
A viral disease, seen worldwide, which targets the bursal component of the immune system of chickens. In addition to the direct economic effects of the clinical disease, the damage caused to the immune system interacts with other pathogens to cause significant effects. The age up to which infection can cause serious immunosuppression varies between 14 and 28 days according to the antigen in question. Generally speaking the earlier the damage occurs the more severe the effects. The infective agent is a Birnavirus (Birnaviridae), Sero-type 1 only, first identified in the USA in 1962. (Turkeys and ducks show infection only, especially with sero-type 2). Morbidity is high with a mortality usually 0–20% but sometimes up to 60%. Signs are most pronounced in birds of 4–6 weeks and White Leghorns are more susceptible than broilers and brown-egg layers. The route of infection is usually oral, but may be via the conjunctiva or respiratory tract, with an incubation period of 2–3 days. The disease is highly contagious. Mealworms and litter mites may harbour the virus for 8 weeks, and affected birds excrete large amounts of virus for about 2 weeks post infection. There is no vertical transmission. The virus is very resistant, persisting for months in houses, faeces etc. Subclinical infection in young chicks results in: deficient immunological response to

Newcastle disease, Marek's disease and Infectious Bronchitis; susceptibility to Inclusion Body Hepatitis and gangrenous dermatitis and increased susceptibility to CRD.

Signs
Depression.
Inappetance.
Unsteady gait.
Huddling under equipment.
Vent pecking.
Diarrhoea with urates in mucus.

Post-mortem lesions
Oedematous bursa (may be slightly enlarged, normal size or reduced in size depending on the stage), may have haemorrhages, rapidly proceeds to atrophy.
Haemorrhages in skeletal muscle (especially on thighs).
Dehydration.
Swollen kidneys with urates.

Diagnosis
Clinical disease – History, lesions, histopathology.
Subclinical disease – A history of chicks with very low levels of maternal antibody (Fewer than 80% positive in the immunodifusion test at day old, Elisa vaccination date prediction < 7 days), subsequent diagnosis of 'immunosuppression diseases' (especially inclusion body hepatitis and gangrenous dermatitis) is highly suggestive. This may be confirmed by demonstrating severe atrophy of the bursa, especially if present prior to 20 days of age. The normal weight of the bursa in broilers is about 0.3% of bodyweight, weights below 0.1% are highly suggestive. Other possible causes of early immunosuppression are severe mycotoxicosis and managment problems leading to severe stress. *Variants*: There have been serious problems with early Gumboro disease in chicks with maternal immunity, especially in the Delmarva Peninsula in the USA. IBD viruses have been isolated and shown to have significant but not complete cross-protection. They are all sero-type 1. Serology: antibodies can be detected as early as 4–7 days after infection

and these last for life. Tests used are mainly Elisa, (previously SN and DID). Half-life of maternally derived antibodies is 3.5–4 days. Vaccination date prediction uses sera taken at day old and a mathematical formula to estimate the age when a target titre appropriate to vaccination will occur.

Differentiate clinical disease from: Infectious bronchitis (renal); Cryptosporidiosis of the bursa (rare); Coccidiosis; Haemorrhagic syndrome.

Treatment

No specific treatment is available. Use of a multivitamin supplement and facilitating access to water may help. Antibiotic medication may be indicated if secondary bacterial infection occurs.

Prevention

Vaccination, including passive protection via breeders, vaccination of progeny depending on virulence and age of challenge. In most countries breeders are immunised with a live vaccine at 6–8 weeks of age and then re-vaccinated with an oil-based inactivated vaccine at 18 weeks. A strong immunity follows field challenge. Immunity after a live vaccine can be poor if maternal antibody was still high at the time of vaccination. When outbreaks do occur, biosecurity measures may be helpful in limiting the spread between sites, and tracing of contacts may indicate sites on which a more rebust vaccination programme is indicated.

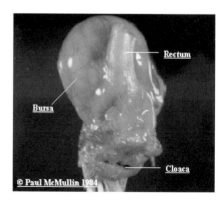

Figure 23. This shows the anatomical relationship between the bursa, the rectum and the vent. This bursa is from an acutely affected broiler. It is enlarged, turgid and oedematous.

Infectious Coryza

Introduction
A usually acute, sometimes chronic, highly infectious disease of chickens, occasionally pheasants and guinea-fowl, characterised by catarrhal inflammation of the upper respiratory tract, especially nasal and sinus mucosae. It is caused by the bacterium *Haemophilus paragallinarum* and is seen in many countries especially in multi-age farms that are never depopulated. Morbidity is high but mortality low if uncomplicated although it may be up to 20%. The route of infection is conjunctival or nasal with an incubation period of 1–3 days followed by rapid onset of disease over a 2–3 day period with the whole flock affected within 10 days, resulting in increased culling. Carriers are important with transmission via exudates and by direct contact. It is not egg transmitted. The bacterium survives 2–3 days outside the bird but is easily killed by heat, drying and disinfectants. Intercurrent respiratory viral and bacterial infections are predisposing factors.

Signs
Facial swelling.
Purulent ocular and nasal discharge.
Swollen wattles.
Sneezing.
Dyspnoea.
Loss in condition.
Drop in egg production of 10–40%.
Inappetance.

Post-mortem lesions
Catarrhal inflammation of nasal passages and sinuses.
Conjunctivitis.
Eye-lid adherence.
Caseous material in conjunctiva/sinus.
Tracheitis.

Diagnosis
A presumptive diagnosis may be made on signs, lesions, identification of the bacteria in a Gram-stained smear from sinus.

Confirmation is by isolation and identification – requires X (Haematin) and V (NAD) factors, preferably in raised CO_2 such as a candle jar. Serology: HI, DID, agglutination and IF have all been used but are not routine. Differentiate from Mycoplasmosis, respiratory viruses, chronic or localised pasteurellosis and vitamin A deficiency.

Treatment

Streptomycin, Dihydrostreptomycin, sulphonamides, tylosin, erythromycin. Flouroquinolones are bactericidal and might prevent carriers.

Prevention

Stock coryza-free birds on an all-in/all-out production policy. Bacterin at intervals if history justifies or if multi-age; at least two doses are required. Commercial bacterins may not fully protect against all field strains but reduce the severity of reactions. Live attenutated strains have been used but are more risky. Controlled exposure has also been practised. Vaccines are used in areas of high incidence. Birds recovered from challenge of one sero-type are resistant to others, while bacterins only protect against homologous strains.

INFECTIOUS LARYNGOTRACHEITIS, ILT

Introduction

A herpesvirus (pathogenicity can vary) infection of chickens, pheasants, peafowl and turkeys with a morbidity of 50–100% and a mortality usually 10–20% but sometimes up to 70%. Recovered and vaccinated birds are long-term carriers. The route of infection is via upper respiratory tract and conjunctiva or possibly oral and the course of the disease is up to 6 weeks. Fairly slow lateral spread occurs in houses. Transmission between farms can occur by airborne particles or fomites. The virus is highly resistant outside host but is susceptible to disinfectants. Movement and mixing of stock and reaching point of lay are predisposing factors.

Signs

Dyspnoea.

Gasping.
Coughing of mucus and blood.
Drop in egg production.
Ocular discharge.
Sinusitis.
Nasal discharge (low pathogenicity strains).

Post-mortem lesions

Severe laryngotracheitis, often with blood in lumen, caseous plugs may be present.
Microscopically – intranuclear inclusions in tracheal epithelium.

Diagnosis

Signs, lesions, in severe form may be enough. Isolation in CE CAMs, histology, IFA, PCR. Differentiate from Newcastle disease, severe bronchitis. Sera may be examined by VN or Elisa.

Treatment

None, antibiotics to control secondary bacterial infection if this is marked.

Prevention

Quarantine, vaccination, if enzootic or epizootic in an area, after 4 weeks of age. All-in/all-out operation. Keep susceptible stock separate from vaccinated or recovered birds. Apply strict biosecurity in moving equipment or materials between these these categories of stock.

INTUSSUSCEPTION

Introduction

A condition of chickens associated with increased intestinal motility. Coccidiosis, worms and other forms of enteritis are predisposing factors.

Signs

Mainly sudden deaths.

Post-mortem lesions
Telescoping of the bowel, usually in the lower small intestine; it may actually protrude at the vent and be cannibalised.

Diagnosis
Typical post-mortem lesions

Treatment
As appropriate for the underlying disease condition.

Prevention
Control of coccidiosis and other causes of intestinal inflammation. Careful attention to feeding practices to avoid anticipation of feed arrival if on restricted feeding.

MALABSORPTION SYNDROME, RUNTING/STUNTING

Introduction
A condition of chickens and turkeys that may be associated with several different viruses, for example enteroviruses, enterovirus-like particles, reoviruses, rotavirus etc. It is suspected that these viruses can be vertically transmitted and the most severe signs in young birds are associated with this. Diarrhoea in older birds may be an effect of on-farm infection. The condition has been seen in Europe, North and South America and Australia. Poor management may contribute to the problem.

Signs
Uneven growth.
Stunting (temporary).
Runting (permanent).
Poor feathering.
Abnormal feathers ('helicopter wings' 'yellow-heads').
Diarrhoea.
Pale shanks in corn- (maize-) fed birds (often associated with orange intestinal contents and/or faeces).
Eating faeces.

Post-mortem lesions
Enteritis.

Pancreatic atrophy/fibrosis and pancreatic atrophy (rather variable).
Pot-bellied appearance.
Sometimes osteomyelitis and/or rickets.

Diagnosis
Pathology, direct electron microscope examination of intestinal contents. Poor early management (feed and water supply, temperature control) may lead to a similar picture in the absence of specific infection.

Treatment
Daily cull of affected birds between 14 and 28 days.

Prevention
Good broiler farm hygiene, good parent nutrition and egg selection and santitation, avoidance of intercurrent disease and management problems (such as chilling).

Figure 24. Intestines of a young broiler chick suffering from malabsorption syndrome. They are distended with poorly digested feed. A sample of the faeces produced is shown at the bottom of the picture – poorly digested food enclosed in mucus.

MAREK'S DISEASE

Introduction
A Herpes virus infection of chickens, and rarely turkeys in close association with chickens, seen worldwide. From the 1980s and

1990s highly virulent strains have become a problem in North America and Europe. The disease has various manifestations: a) Neurological – Acute infiltration of the CNS and nerves resulting in 'floppy broiler syndrome' and transient paralysis, as well as more long-standing paralysis of legs or wings and eye lesions; b) Visceral – Tumours in heart, ovary, tests, muscles, lungs; c) Cutaneous – Tumours of feather follicles.

Morbidity is 10–50% and mortality up to 100%. Mortality in an affected flock typically continues at a moderate or high rate for quite a few weeks. In 'late' Marek's the mortality can extend to 40 weeks of age. Affected birds are more susceptible to other diseases, both parasitic and bacterial. The route of infection is usually respiratory and the disease is highly contagious being spread by infective feather-follicle dander, fomites, etc. Infected birds remain viraemic for life. Vertical transmission is not considered to be important. The virus survives at ambient temperature for a long time (65 weeks) when cell associated and is resistant to some disinfectants (quaternary ammonium and phenol). It is inactivated rapidly when frozen and thawed.

Signs
Paralysis of legs, wings and neck.
Loss of weight.
Grey iris or irregular pupil.
Vision impairment.
Skin around feather follicles raised and roughened.

Post-mortem lesions
Grey-white foci of neoplastic tissue in liver, spleen, kidney, lung, gonads, heart, and skeletal muscle.
Thickening of nerve trunks and loss of striation.
Microscopically – lymphoid infiltration is polymorphic.

Diagnosis
History, clinical signs, distribution of lesions, age affected, histopathology.
Differentiate from Lymphoid leukosis, botulism, deficiency of thiamine, deficiency of Ca/Phosphorus/Vitamin D, especially at the start of lay.

Treatment
None.

Prevention
Hygiene, all-in/all-out production, resistant strains, vaccination generally with 1500 PFU of HVT at day old (but increasingly by in-ovo application at transfer), association with other strains (SB1 Sero-type 2) and Rispen's. It is common practice to use combinations of the different vaccine types in an effort to broaden the protection achieved. Genetics can help by increasing the frequency of the B21 gene that confers increased resistance to Marek's disease challenge.

Mycoplasma gallisepticum infection, M.G., Chronic Respiratory Disease – Chickens

Introduction
Infection with *Mycoplasma gallisepticum* is associated with slow onset, chronic respiratory disease in chickens, turkeys, game birds, pigeons and other wild birds. Ducks and geese can become infected when held with infected chickens. In turkeys it is most associated with severe sinusitis (see separate description in the turkey section). The condition occurs worldwide, though in some countries this infection is now rare in commercial poultry.In others it is actually increasing because of more birds in extensive production systems that expose them more to wild birds. In adult birds, though infection rates are high, morbidity may be minimal and mortality varies. The route of infection is via the conjunctiva or upper respiratory tract with an incubation period of 6–10 days. Transmission may be transovarian, or by direct contact with birds, exudates, aerosols, airborne dust and feathers, and to a lesser extent fomites. Spread is slow between houses and pens suggesting that aerosols are not normally a major route of transmission. Fomites appear to a significant factor in transmission between farms. Recovered birds remain infected for life; subsequent stress may cause recurrence of disease. The infectious agent survives for only a matter of days outwith birds although prolonged survival has been reported in egg yolk and allantoic fluid, and in lyophilised material. Survival

seems to be improved on hair and feathers. Intercurrent infection with respiratory viruses (IB, ND, ART), virulent *E. coli*, *Pasteurella* spp. *Haemophilus*, and inadequate environmental conditions are predisposing factors for clinical disease.

Signs
Coughing.
Nasal and ocular discharge.
Poor productivity.
Slow growth.
Leg problems.
Stunting.
Inappetance.
Reduced hatchability and chick viability.
Occasional encephalopathy and abnormal feathers.

Post-mortem lesions
Airsacculitis.
Pericarditis.
Perihepatitis (especially with secondary *E. coli* infection).
Catarrhal inflammation of nasal passages, sinuses, trachea and bronchi.
Occasionally arthritis, tenosynovitis and salpingitis in chickens.

Diagnosis
Lesions, serology, isolation and identification of organism, demonstration of specific DNA (commercial PCR kit available). Culture requires inoculation in mycoplasma-free embryos or, more commonly in Mycoplasma Broth followed by plating out on Mycoplasma Agar. Suspect colonies may be identified by immuno-flourescence. Serology: serum agglutination is the standard screening test, suspect reactions are examined further by heat inactivation and/or dilution. Elisa is accepted as the primary screening test in some countries. HI may be used, generally as a confirmatory test. Suspect flocks should be re-sampled after 2–3 weeks. Some inactivated vaccines for other diseases induce 'false positives' in serological testing for 3–8 weeks. PCR is possible if it is urgent to determine the flock status. Differentiate from Infectious Coryza, Aspergillosis, viral respiratory diseases, vitamin A deficiency, other *Mycoplasma*

infections such as *M. synoviae* and *M. meleagridis* (turkeys).

Treatment
Tilmicosin, tylosin, spiramycin, tetracyclines, fluoroquinolones. Effort should be made to reduce dust and secondary infections.

Prevention
Eradication of this infection has been the central objective of official poultry health programmes in most countries, therefore M.g. infection status is important for trade in birds, hatching eggs and chicks. These programmes are based on purchase of uninfected chicks, all-in/all-out production, biosecurity, and routine serological monitoring. In some circumstances preventative medication of known infected flocks may be of benefit. Live attenuated or naturally mild strains are used in some countries and may be helpful in gradually displacing field strains on multi-age sites. Productivity in challenged and vaccinated birds is not as good as in M.g.-free stock.

MYCOPLASMA SYNOVIAE INFECTION, M.S.
INFECTIOUS SYNOVITIS

Introduction
Infection with *Mycoplasma synoviae* may be seen in chickens and turkeys in association with synovitis and/or airsacculitis. It occurs in most poultry-producing countries, especially in commercial layer flocks. Infection rates may be very high. Spread is generally rapid within and between houses on a farm, whilst illness is variable and mortality less than 10%. Infection is via the conjunctiva or upper respiratory tract with a long incubation period, 11–21 days following contact exposure. Transmission may be transovarian, or lateral via respiratory aerosols and direct contact. Survival of the infectious agent outwith the bird is poor but fomite transmission between farms is important. Predisposing factors include stress and viral respiratory infections.

Signs
There may be no signs.
Depression.

Inappetance.
Ruffled feathers.
Lameness.
Swelling of hocks, shanks and feet (sometimes severe and bilaterally asymmetrical).
Faeces may be green in acute infections.
Effects on egg production appear to be minor under good management.

Post-mortem lesions

Joints and tendon sheaths have viscid grey to yellow exudate.
Some strains can lead to amyloidosis.
Swollen liver, spleen and kidney have been seen in the past but are not common now.
Green liver.
Exudate becomes caseous later.
Sternal bursitis.
Airsacculitis – usually in heavy broilers and associated with condemnations.

Diagnosis

Lesions, serology, isolation (difficult – requires NAD) and identification. Differentiate from viral arthritis, staphylococcal arthritis, *Mycoplasma gallisepticum* infections, *Ornithobacterium rhinotracheale*, viral respiratory disease with colibacillosis. Serology: SAG used routinely, Elisa in some countries – PCR and/or culture used to confirm. False positives post inactivated vaccines are, if anything more common than in the case of M.g.

Treatment

Tilmicosin, chlortetracycline, oxytetracycline, tylosin.

Prevention

Eradication of this infection is also possible using similar techniques as described for *Mycoplasma gallisepticum*. These are based on purchase of uninfected chicks, all-in/all-out production, and biosecurity. Maintenance of *Mycoplasma synoviae* free status seems to be more difficult than for *Mycoplasma gallisepticum*. In some circumstances preventative medication of known infected flocks may be of benefit. Vaccines are not widely used though they are available in some countries. Infected birds do develop some immunity to the effects of repeated inoculation.

MYCOTOXICOSIS

Introduction

Mycotoxicosis refers to all of those diseases caused by the effects of toxins produced by moulds. Disease is often subclinical and may be difficult to diagnose. Problems occur worldwide, but especially climates with high temperature and humidity and where grain is harvested with high water content. Economic impact is considerable in some countries. A number of different types are recognised: aflatoxins are produced by *Aspergillus flavus*; T2 fusariotoxins by *Fusarium* spp. (mouth lesions and thin eggshells); ochratoxins by *Aspergillus ochraceus* (interferes with functions of kidney, proventriculus and gizzard); rubratoxin by *Penicillium rubrum* (interferes with thiamine metabolism and causes symptoms of deficiency). Other mycotoxins certainly occur. Mortality is variable but all are detrimental to bird health and are resistant to heat inactivation. The following species may be affected, in decreasing order of susceptibility: ducks, turkeys, geese, pheasants, chickens. The route of infection is by ingestion of fungal spores, which are readily carried in the air. High grain humidity, and damage due to insects, as well as poor storage conditions are major predisposing causes. Once toxins have been formed it is difficult to avoid their biological effects; they also increase susceptibility to bacterial diseases. Both fungal spores and formed toxins are generally highly resistant. Affected flocks return to normal mortality by 7 to 15 days after removal of the toxins. Some believe that mycotoxicosis is an important factor in fatty liver syndrome. Aflatoxins are known to inibit the synthesis and transport of lipids in the liver. Deficiencies of fat-soluble vitamins (A, D, E, and K) are also sometimes seen in aflatoxicosis. Multiplication of moulds in cereals requires selenium and this element is also important for the production of hepatic lipases. Aflatoxins have been shown to be carcinogenic in rodents so there may be public health issues relating to the effective control of these problems.

Signs

Signs vary with the species affected, the mycotoxin, the dose ingested and the period of exposure.

Diarrhoea.

Paralysis or incoordination.

Reduced feed efficiency.

Reduced weight gain or egg production/hatchability.

Increased condemnations.

Pale shanks, combs, bone marrows.

Post-mortem lesions

Lesions also vary in accorance with the same factors as signs. Mycotoxins can cause damage to mucosae with which they come in contact.

They can also be absorbed and affect blood coagulation, resulting in petechiae and larger haemorrhages in various tissues.

Liver and kidney lesions – livers may be enlarged and fatty or show bile retention or tumours.

Enteritis of variable degree may be seen.

Hydropericardium.

Pale bone marrow.

Regression of the bursa of Fabricius.

Gizzard erosions.

Diagnosis

In severe cases a presumptive diagnosis may be based on the history, signs and lesions. Histology may be beneficial in some cases, as may identification and quantification of toxins in samples of feed or feed residue. Differentiate from poor nutrition, poor management, physical damage to tissues, and infectious bursal disease.

Treatment

The most effective treatment is removal of the source of toxins. Addition of antifungal feed preservatives is also helpful. Increasing protein level in the feed until mortality reduces may also be beneficial. Administration of soluble vitamins and selenium (0.2 ppm), along with finely divided copper sulphate in the feed 1kg/ton for 7 days (where approved) has been used.

Prevention

Mycotoxicoses may be prevented by careful choice of feed raw materials, reduction in water content of the raw materials and hygienic storage. Antimycotic feed additives may also be used but may not deal with toxins already formed. Feeds with high levels of fishmeals are particularly susceptible and should not be stored for more than 3 weeks. Pelletising feed may reduce fungal counts but does not affect toxins. Certain minerals additives have been shown to bind mycotoxins and reduce their effects. Good stock control, management of feeders and bins, and avoidance of feed spillage are all important.

NECROTIC ENTERITIS

Introduction

An acute or chronic enterotoxemia seen in chickens, turkeys and ducks worldwide, caused by *Clostridium perfringens* and characterised by a fibrino-necrotic enteritis, usually of the mid–small intestine. Mortality may be 5–50%, usually around 10%. Infection occurs by faecal–oral transmission. Spores of the causative organism are highly resistant. Predisposing factors include coccidiosis/coccidiasis, diet (high protein), in ducks possibly heavy strains, high viscosity diets (often associated with high rye and wheat inclusions in the diet), contaminated feed and/or water, other debilitating diseases.

Signs

Depression.
Ruffled feathers.
Inappetance.
Closed eyes.
Immobility.
Dark coloured diarrhoea.
Sudden death in good condition (ducks).

Post-mortem lesions

Small intestine (usually middle to distal) thickened and distended.
Intestinal mucosa with diptheritic membrane.

Intestinal contents may be dark brown with necrotic material. Reflux of bile-stained liquid in the crop if upper small intestine affected.

Affected birds tend to be dehydrated and to undergo rapid putrefaction.

Diagnosis

A presumptive diagnosis may be made based on flock history and gross lesions Confirmation is on the observation of abundant rods in smears from affected tissues and a good response to specific medication, usually in less than 48 hours.

Treatment

Penicillins (e.g. phenoxymethyl penicillin, amoxycillin), in drinking water, or Bacitracin in feed (e.g. 100 ppm). Treatment of ducks is not very successful, neomycin and erythromycin are used in the USA. Water medication for 3–5 days and in-feed medication for 5–7 days depending on the severity.

Prevention

Penicillin in feed is preventive, high levels of most growth promotors and normal levels of ionophore anticoccidials also help. Probiotics may limit multiplication of bacteria and toxin production. In many countries local regulations or market conditions prevent the routine use of many of these options.

Figure 25. Severe lesions of necrotic enteritis affecting the small intestine of broilers. The sample at the bottom of the picture is still focal, the upper has formed a thick crust of necrotic material.

Non-starter and 'Starve-outs'

Introduction
This is a condition seen worldwide in young chickens and turkeys due to failure to begin normal food consumption. Morbidity is up to 10% and mortality close to 100%. It is commonest in progeny of young parent flocks (relatively small yolks) or where managers are inexperienced, and is associated with difficulty maintaining brooding temperature and poor management of feeders and drinkers. Acute viral infections or heavily contaminated drinking water may have a similar effect to bad management in reducing feed intake. Mortality peaks at 3–5 days for birds that fail to eat at all, whereas it tends to peak between 6–10 days for chicks that begin eating but then cease.

Signs
Failure to grow.

Post-mortem lesions
Crop empty.
Most organs normal.
Poor development.
Gall bladder distended, bile staining of adjacent tissues.
Gizzard may be impacted with litter.

Diagnosis
Based on post-mortem lesions. Differentiate from omphalitis/yolk sac infection, aspergillosis.

Treatment
Correction of management problems, soluble multivitamin supplementation may help.

Prevention
Review brooding management, nutrition of young parents, age of collection and weight of hatching eggs, good drinking water hygiene, good hygiene and biosecurity in brooding areas to minimise early disease challenges.

'Oregon Disease', Deep Pectoral Myopathy

Introduction
This condition is so named because it was first identified in adult breeding turkeys in Oregon State. It has since been seen in turkeys, in broiler parent chickens and also in large broiler chickens in various places. It does not usually cause any mortality or obvious clinical signs and so it is usually identified after slaughter. It is caused by a reduction in the blood supply to the deep pectoral muscles. Genetics may play a role in that it has been suggested that blood supply to the affected muscles is reduced in some heavy meat type birds. The condition can be reproduced by causing intense exercise of this muscle. Because it is enclosed in a relatively unyielding membrane, any swelling of the muscle tends to cut off the blood supply. Without adequate blood supply the tissue of the muscle begins to die, or suffer necrosis.

Signs
None.

Post-mortem lesions
Acute or chronic necrosis of the deep pectoral muscle on one or both sides.

If recent, the muscle may be swollen and pale, with oedema within it and on its surface. If the condition is of over 7 days duration the muscle is dry and often shows greenish tinges. It may also start to be enclosed in a fibrous capsule. If of very long duration, it may become a healed scar.

Diagnosis
Lesions. It is very difficult to detect affected carcases in standard meat inspection. Transillumination of the breast may help identify affected carcases and may be worthwhile if the incidence is high.

Treatment
Not applicable.

Prevention
Avoidance of physical damage of this muscle, and, in particular

excessive exercise. This will normally be due to excessive flapping in association with management activities (e.g. thinning operations in meat birds, artificial insemination in breeding turkeys, weighing birds etc). If inactivated vaccines are administered into the breast muscle it is preferable that they go into the superficial muscle which is better able to cope with swelling.

Figure 26. A typical case of the chronic stages deep pectoral myopathy in a broiler chicken. The tissue in the centre is dry, greenish yellow, and flaking.

ORNITHOBACTERIUM INFECTION, ORT

Introduction

An infection of chickens and turkeys with the bacterium *Ornithobacterium rhinotracheale* (ORT). This was first identified as a new disease syndrome of turkeys in Germany in the early 1990s. The slow-growing bacterium was named in 1994. It had been previously isolated in a number of other countries. The infection is common in chickens and turkeys. The severity of its effects depends on the pathogenicity of the particular strain and other risk factors such as viral infections, ventilation problems, age at infection etc. Inoculation of fresh material from a case can reproduce lung lesions similar to those caused by *Pasteurella multocida* infection in previously uninfected turkeys over 10 weeks of age. It commonly exacerbates respiratory disease caused by pneumovirus infections in turkeys. In broiler chickens its main importance seems to be as a cause of airsacculitis in apparently healthy flocks that is only identified at slaughter. This can cause significant losses through condemnations. Important cofactors may include respiratory viral vaccines (Newcastle disease, Infectious Bronchitis), field challenge with respiratory viruses, *Bordetella avium* infection, and *E. coli* infections. Some uncertainty exists about mechanisms of transmission. Within the poultry house it is likely to be by aerosols, direct contact and drinkers. There is some evidence that hatcheries may play a role in the epidemiology, perhaps through survival of the organism on shell surfaces or shell membranes. A range of sero-types have been identified. The range occurring in turkeys is wider than that seen in chickens.

Signs
Coughing, sneezing.
Reduced weight gain.
Reduced egg production.

Post-mortem lesions
Airsacculitis, tracheitis, severe bronchopneumonia.

Diagnosis
Clinical signs and lesions. Confirmation is by isolation of the organism. Cultures from the trachea of birds showing typical signs are preferred. The organism grows slowly producing tiny colonies on blood agar. Growth is more rapid and consistent in an anaerobic jar. *E. coli* overgrowth may occur. It is a Gram-negative pleomorphic organism. Most isolates are oxidase positive and galactosidase positive Serology using Elisa tests has been used, preferably as a flock test comparing results before and after the challenge.

Treatment
The sensitivity of ORT to antibiotic is highly variable. Initially in Germany most strains were sensitive to amoxycillin, chlortetracycline but not to enrofloxacin. Amoxycillin sensitivity remains good, also most are sensitive to tiamulin. In France and Belgium most strains were sensitive to enrofloxacin. In the USA in 1998 all strains were sensitive to penicillins and many other microbials, though 54% were sensitive to tetracycline,

neomycin and enrofloxacin. Routine treatment – amoxycillin at 250 ppm in water for 3–7 days or chlortetracycline at 500 ppm for 4–5 days. Tilmicosin is also effective at 10–20mg/kg and should be used in the early stages of the disease.

Prevention

This is based on good hygiene, therapy and vaccination. Hygiene – it is very sensitive *in vitro* to a range of chemicals. ORT is a major problem on multi-age sites. Preventative medication with the products listed in the previous section may be beneficial in some circumstances. Vaccination – there are issues relating to availability, regulation, cost etc. Some work has been done on live vaccine in the USA. An inactivated vaccine is licensed for broiler parents in Europe; it requires two doses. Some vaccines seem to have difficulty in producing a long-lasting serological response to ORT. Vaccination of breeder flocks to protect progeny has shown benefits in chickens but, because of the age of slaughter, this is unlikely to work in turkeys. Satisfactory disease control depends on good management and biosecurity.

Figure 27. Severely consolidated lung from an acute ORT infection in 14-week-old turkey growers. The lung is solid and covered by pus/ purulent exudate. Similar lesions may be found in Pasteurellosis.

OSTEOPOROSIS, CAGE FATIGUE

Introduction
A condition of chickens, turkeys and ducks seen worldwide and due to Vitamin D or calcium deficiency. Predisposing factors include small body size, high production.

Signs
Lameness.
Soft bones and beak.
Birds go off legs.
Drop in production.
Soft-shelled eggs.
Enlarged hocks.
Birds rest squatting.

Post-mortem lesions
Bones soft and rubbery.
Epiphyses of long bones enlarged.
Beading and fracture of ribs.
Beak soft.
Parathyroids enlarged.

Diagnosis
History, signs, lesions, bone ash. Differentiate from Marek's disease, spondylitis.

Treatment
Over-correct ration vitamin D, calcium carbonate capsules, place on litter.

Prevention
Supplementation of vitamin D, proper Ca and P levels and ratio, antioxidants. As birds progressively mobilise skeletal calcium for egg production through lay, skeletal size and mineral content at point of lay are critical.

Paramyxovirus 1 or Newcastle disease

Introduction

A highly contagious viral disease affecting poultry of all ages. Affected species include chickens, turkeys, pigeons and ducks. The condition is rarely diagnosed in ducks but is a possible cause of production drops/fertility problems. Other species can be infected including mammals occasionally (e.g. conjunctivitis in man). The virus involved is Paramyxovirus PMV-1, which is of variable pathogenicity. Signs are typically of disease of the nervous, respiratory or reproductive systems. Morbidity is usually high and mortality varies 0–100%. Higher mortality is seen in velogenic disease in unvaccinated stock.

Four manifestations have been identified:

ND – Velogenic Viscerotropic (VVND) – sometimes called 'asiatic' or exotic. It is highly virulent for chickens, less for turkeys and relatively apathogenic in psittacines.

ND – Neurotropic Velogenic – Acute and fatal in chickens of any age causing neurological and some respiratory signs. Intestinal lesions are absent.

ND – Mesogenic – Mortality and nervous signs in adult. These viruses have sometimes been used as vaccines in previously immunised birds.

ND – Lentogenic – Mild disease, sometimes subclinical. Can affect any age. Strains can be developed as vaccines.

Transmission is via aerosols, birds, fomites, visitors and imported psittacines (often asymptomatic). It is not usually vertical (but chicks may become infected in hatcheries from contaminated shells). The virus survives for long periods at ambient temperature, especially in faeces and can persist in houses (in faeces, dust etc). for up to 12 months. However it is quite sensitive to disinfectants, fumigants and sunlight. It is inactivated by temperatures of 56°C for 3 hours or 60°C for 30 min, acid pH, formalin and phenol, and is ether sensitive.

Signs

Signs are highly variable and will depend on the nature of the infecting virus (see above), the infective dose and the degree of immunity from previous exposure or vaccination.

Sudden Death
Depression.
Inappetance.
Coughing.
Dyspnoea.
Diarrhoea.
Nervous signs.
Paralysis.
Twisted neck.
Severe drop in egg production.
Moult.

Post-mortem lesions

Airsacculitis.
Tracheitis.
Necrotic plaques in proventriculus, intestine, caecal tonsil.
Haemorrhage in proventriculus.
Intestinal lesions primarily occur in the viscerotropic form.

Diagnosis

A presumptive diagnosis may be made on signs, post-mortem lesions, rising titre in serology. It is confirmed by isolation in CE, HA+, HI with ND serum or DID (less cross reactions), IFA. Cross-reactions have mainly been with PMV-3. Pathogenicity evaluated by Mean Death Time in embryos, intracerebral or IV pathogenicity in chicks. Samples – tracheal or cloacal. Differentiate from Infectious bronchitis, laryngotracheitis, infectious coryza, avian influenza, EDS-76, haemorrhagic disease, encephalomyelitis, encephalomalacia, intoxications, middle ear infection/skull osteitis, pneumovirus infection.

Treatment

None, antibiotics to control secondary bacteria.

Prevention

Quarantine, biosecurity, all-in/all-out production, vaccination. It is common to monitor response to vaccination, especially in breeding birds by the use of routine serological monitoring. HI has been used extensively; Elisa is now also used. These tests do not directly evaluate mucosal immunity, however. Vaccination programmes should use vaccines of high potency, which are adequately stored and take into account the local conditions. A typical programme may involve Hitchner B1 vaccine at day old followed by LaSota-type vaccine at 14 days. The LaSota-type vaccine may even repeated at 35–40 days of age if risk is high. Use of spray application is recommended but it needs to be applied with care to achieve good protection with minimal reaction. Inactivated vaccines have largely replaced the use of live vaccines in lay but they do not prevent local infections. To prevent or reduce vaccinal reactions in young chicks it is important that day old have uniform titres of maternal immunity. Vaccinal reactions may present as conjunctivitis, snicking, and occasionally gasping due to a plug of pus in the lower trachea. In some countries it has been customary to provide antibiotics prophylactically during periods of anticipated vaccinal reaction. Use of *Mycoplasma gallisepticum*-free stock under good management reduces the risk of vaccinal reactions.

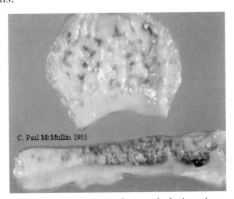

Figure 28. Severe haemorrhagic and necrotic lesions in proventriculus and Peyers patches in the intestines of a broiler chicken suffering from one of the severe forms of Newcastle disease (viscerotropic velogenic).

Paramyxovirus 2 - Yucaipa Disease

Introduction
An infection of chickens, turkeys and passerine cage birds with Paramyxovirus PMV-2 results in disease with variable mortality, depending on the species affected and whether infection is complicated by other factors and diseases. In chickens it is usually mild, whereas turkeys are more severely affected. Transmission is by wild birds and fomites. Wild birds are believed to be especially important. Vertical transmission does not usually occur.

Signs
Depression.
Inappetance.
Coughing.
Drop in egg production.

Post-mortem lesions
Not characteristic.

Diagnosis
Isolation in CE, HA+, HI with ND serum or DID (less cross reactions), IFA. Differentiate from Infectious bronchitis, laryngotracheitis, EDS-76, haemorrhagic disease.

Treatment
No specific treatment, antibiotics to control secondary bacteria.

Prevention
Quarantine, biosecurity, all-in/all-out production.

Pendulous crop

Introduction
A complex condition of turkeys and chickens associated with coarse fibrous feed. Over-distension of crop in young birds because of an interruption in feed supply may predispose.

Signs
Enlargement of crop.

Post-mortem lesions
Crop distended with feed, may be impacted, have ulcerations and sometimes mycosis.
Crop contents semi-liquid and foul smelling.
Lack of muscle tone.

Diagnosis
Signs, lesions.

Treatment
None.

Prevention
Remove predisposing factors, if these can be identified.

PROVENTRICULAR WORMS

Introduction
Dispharynx, *Tetrameres* and *Cyrnea* are nematodes, spiruroid worm parasites of poultry and game birds. They are found in the mucosa of crop and proventriculus causing mainly inapparent infections. Infection is by the oral route on exposure to the intermediate hosts, including crustaceans, grasshoppers and cockroaches.

Signs
Anaemia.
Diarrhoea.
Emaciation in heavily infected young chicks (*Tetrameres*).

Post-mortem lesions
Inflammation, ulceration, haemorrhage, necrosis, and thickening of mucosa of proventriculus.

Diagnosis
Macroscopic examination of lesions, identification of worms.

Treatment
Not usually required.

Prevention
Prevention of access to intermediate hosts.

Pullet Disease, Bluecomb, Avian Monocytosis

Introduction
A sudden onset condition of chickens early in lay with high morbidity and a mortality of 0–50%. It is associated with hot weather, water deprivation, toxin and possibly a virus infection. Affected birds have an increase in circulating monocytes in their blood. It was commonly reported prior to 1960 but is now rare.

Signs
- Depression.
- Loss of appetite.
- Crop distension.
- Dark comb and wattles.
- Watery diarrhoea.
- Dehydration.
- Drop in egg production.

Post-mortem lesions
- Dehydration.
- Skin cyanosis of head.
- Skeletal muscle necrosis.
- Crop distended.
- Catarrhal enteritis, mucoid casts in intestine.
- Liver swollen with foci.
- Kidneys swollen.
- Pancreas chalky.

Diagnosis
History, lesions, elimination of other causes. Differentiate from fowl cholera, vibriosis, capillariasis, coccidiosis, IBD and typhoid.

Treatment
Adequate water supply, antibiotics, molasses, multivitamins in water.

Prevention
Good management, hygiene, diet and water supply.

Chapter 6 - Diseases and Syndromes: Chickens & Various Species

RED MITE AND NORTHERN FOWL MITE

Introduction
Arachnid mites, which are external parasites of chickens and turkeys. *Dermanyssus gallinae,* the Common Red Mite, feeds by sucking blood, mainly at night and may transmit fowl cholera and other diseases. *Ornithonyssus bursae,* the Northern Fowl Mite spends its entire life cycle on the bird and can multiply more rapidly as a result.

Signs
Presence of grey to red mites up to 0.7 mm.
Birds restless.
May cause anaemia and death in young birds. Loss of condition.
Pale comb and wattles.
Drop in egg production.
Spots on eggs.
Staff complaints – itching.

Post-mortem lesions
Anaemia.

Diagnosis
Number and type of mites identified. However keep in mind that the majority of the population of *Dermanyssus* is in the environment so it is necessary to monitor infection levels on feeder tracks, in nest boxes, cracks, crevices etc.

Treatment
Pyrethroids, organophosphates, carbamates, citrus extracts, vegetable oil and mineral-based products (both liquid sand dusts) have been used to control red mites in the environment. For northern fowl mite it is essential to apply approved insecticides to the affected birds.

Prevention
Thorough cleaning, fumigation and insecticide treatment at turn-around. Filling of cracks and crevices, and good design of new equipment to limit harbourages for red mites. Effective monitoring of mite numbers and implementation of control measures before birds are heavily infested improves control.

Respiratory Adenovirus Infection, 'Mild Respiratory Disease'

Introduction
An adenovirus infection of chickens with a morbidity of 1–10% and a mortality of 1–10%; at least 12 sero-types have been described and these may be isolated from healthy chickens. Infected birds may remain carriers for a few weeks. Transmission may be vertical and lateral, and by fomites. The virus is generally resistant to disinfectants (ether, chloroform, pH), temperature, formaldehyde and iodides work better. Opinions vary as to whether adenovirus can be characterised as a primary respiratory pathogen. It may occur as an exacerbating factor in other types of respiratory disease.

Signs
Mild snick and cough without mortality.

Post-mortem lesions
Mild catarrhal tracheitis.

Diagnosis
History, lesions, intranuclear inclusions in liver. The virus grows well in tissue culture (CE kidney, CE liver).

Treatment
None.

Prevention
Quarantine and good sanitary precautions, prevention of immunosuppression.

Respiratory Disease Complex

Introduction
Respiratory infections in chicken and turkeys are seen worldwide but especially in temperate poultry-producing areas in winter

months. A number of respiratory viruses (Infectious Bronchitis, Avian pneumovirus, Lentogenic Newcastle disease virus, vaccinal and field strains) and bacteria (*Ornithobacterium rhinotracheale, E. coli*) may be involved. Dust, ammonia and other gases, and other factors associated with poor ventilation, may act as predisposing factors. Morbidity is typically 10–20%, mortality 5–10%. If condemned birds are included mortality may be more than 10%.

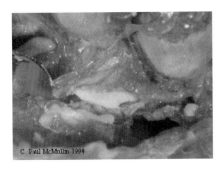

Figure 29. Severe airsacculitis of thoracic air sacs in a broiler close to slaughter. The air sacs are thickened and congested, and have been cut into to reveal a cheesy (caseous) mass of pus. There is also percarditis evident (at top right).

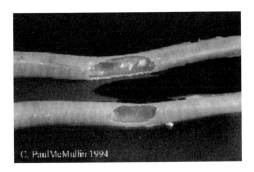

Figure 30. Severe tracheitis is broilers with respiratory disease complex. Both tracheas were congested and the upper sample had mucopurulent material in the lumen.

Signs
Snick.
Sneezing.
Head swelling.
Conjunctivitis.
Nasal exudate.
Rattling noises.

Post-mortem lesions
Severe tracheitis with variable exudate – catarrhal to purulent.
Airsacculitis.
Pericarditis.

Diagnosis
Lesions, serology, response to environmental changes. Differentiate from Chronic Respiratory Disease (Mycoplasmosis).Given that many flocks are vaccinated it is necessary to establish normal serological response in vaccinated flocks in the absence of disease (some of which may, of course, be challenged by some of these pathogens).

Treatment
Antimicrobial treatment of specific bacterial infections.

Prevention
Effective ventilation, sanitation of drinking water, carefully applied appropriate viral vaccines.

Reticuloendotheliosis, Lymphoid tumour disease

Introduction
A retrovirus infection of chickens, turkeys, ducks, geese, and quail with morbidity up to 25%, usually higher in females than males. A gradual increase in mortality and poor growth in broilers may be the main signs. Transmission is lateral and possibly transovarian, and the infection can be spread by mosquitoes or in contaminated

Marek's disease vaccines. There is an incubation period of 5–15 days.

Signs
There may be no obvious signs.
Diarrhoea.
Leg weakness.
Marked stunting when Marek's disease vaccine is contaminated.

Post-mortem lesions
Neoplastic lesions in liver, spleen and kidney.
Sometimes nodules in intestine and caecae.
Sometimes enteritis and proventriculitis (especially in broilers infected with contaminated Marek's disease vaccine).

Diagnosis
Pathology, chick inoculation.

Treatment
None.

Prevention
Formal control programmes have not been documented as the condition is sporadic and relatively self-limiting. Avoidance of contamination of live vaccines is the main control measure practised.

RICKETS (HYPOCALCAEMIC)

Introduction
Vitamin D deficiency or phosphorus/calcium imbalance is seen in chickens, turkeys and ducks worldwide.

Signs
Lameness.
Hock swelling.
Soft bones and beak.
Birds go off legs.
Poor growth.
Birds rest squatting.
Reduction in bodyweight.

Post-mortem lesions
Bones soft and rubbery.
Epiphyses of long bones enlarged.
Beading and fracture of ribs.
Growth plates widened and disorganised.
Beak soft.
Parathyroids enlarged.

Diagnosis
History, signs, lesions. Differentiate from Encephalomalacia, Femoral Head Necrosis.

Treatment
Over-correct ration with three times vitamin D for 2 weeks, or Vitamin D or 25-hydroxy vitamin D in drinking water.

Prevention
Supplementation of vitamin D, proper calcium and phosphorus levels and ratio, antioxidants.

RICKETS (HYPOPHOSPHATAEMIC)

Introduction
Phosphorus deficiency in chickens, turkeys and duck is seen worldwide. Some countries have regulations governing the use of phosphorus in fertiliser with a view to reducing water pollution. This has led feed manufacturers to try various approaches to reduce added phosphorus in diets. Intestinal diseases may reduce absorption. Even subclinical levels of this condition can predispose to Femoral Head Necrosis.

Signs
Lameness.
Hock swelling.
Soft bones and beak.
Birds go off legs.
Poor growth.
Enlarged hocks.
Birds rest squatting.
Reduction in bodyweight.

Post-mortem lesions
Bones soft and rubbery.
Epiphyses of long bones enlarged.
Beading and fracture of ribs.
Growth plates widened and disorganised.
Beak soft.
Parathyroids enlarged.

Diagnosis
History, signs, lesions (analyse Vitamin A or manganese as a guide to premix inclusion).

Treatment
Over-correct ration with 3 times vitamin D for 2 weeks, or vitamin D in drinking water.

Prevention
Supplementation of Vitamin D, proper calcium and phosphorus levels and ratio, antioxidants.

ROTAVIRUS INFECTION

Introduction
Rotavirus infection is seen in chickens, turkeys, guinea fowl, pheasants, partridges and pigeons. The route of infection is oral and maximum viral excretion occurs 2–5 days post infection. It is not known if vertical transmission occurs though virus has been occasionally isolated from embryonated eggs.

Signs
Diarrhoea.

Post-mortem lesions
Viruses replicate mainly in the mature villous epithelial cells.

Diagnosis
Demonstration of virus (PAGE, electron microscopy or Elisa on intestinal contents – submit 6 x .5 g faeces). Virus may be found in asymptomatic birds.

Treatment

No specific treatment for this infection. Control of secondary bacterial enteritis may require antimicrobial medication. Use of a good electrolyte solution is beneficial in the acute stage of infection.

Prevention

Good hygiene in brooding areas.

ROUNDWORM, LARGE – ASCARIDIA

Introduction

Ascaridia sp. are nematode worm parasites, stout white worms up to 12 cms in length, seen worldwide. The parasite species vary: *A. galli* in fowl; *A. dissimilis* in turkeys; and *A. columbae* in pigeons. The route of infection is oral usually by direct ingestion of the embryonated egg and there is a 5–10-week prepatent period, shorter in young birds. The life cycle is similar to that illustrated in the section on the caecal worm *Heterakis*, except that the adults reside in the small intestine and earthworms are not significant paratenic or transport hosts. Adult birds can tolerate burdens asymptomatically. The parasites may be transported by grasshoppers and earthworms and are resistant to environmental effects.

Signs

Loss of condition.
Poor growth.
Listlessness.
Diarrhoea.
Wasting, mainly in young birds.

Post-mortem lesions

Enteritis.
Worms up to 12 cm in length in duodenum and ileum.

Diagnosis

Macroscopic examination with identification of worms, oval smooth-shelled non-embryonated eggs in faeces.

Treatment
Flubendazole, levamisole, piperazine as locally approved.

Prevention
Prevention of contamination of feeders and drinkers with faeces; pasture rotation and regular treatment, especially for young birds.

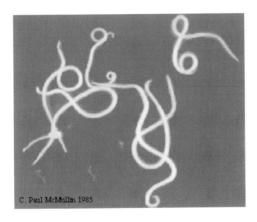

Figure 31. A collection of mature roundworms (*Ascaridia galli*) recovered from the small intestine of a layer chicken in rear. In the bottom left quadrant there are also some immature worms.

RUPTURED GASTROCNEMIUS TENDON

Introduction
This is a condition causing severe lameness in roaster type meat chickens and more rarely in turkeys. Reoviral or staphylococcal arthritis or tenosynovitis may predispose but many cases have no evidence of an inflammatory response. Broiler chickens have an avascular area of tendon just above the hock that is the usual site of rupture. Ruptures of ligaments and joints are also occasionally seen.

Signs
Severe lameness.
The bird may have the hock dropped to the floor. If there is a

greenish tinge to the area of swelling it occurred a few days previously.

Post-mortem lesions
The fresh lesion contains a haematoma with the end of the ruptured tendon within or adjacent to it.

Diagnosis
Signs and lesions are obvious. Histology may be helpful in determining if there is an underlying inflammatory process.

Treatment
None. Affected birds should be culled humanely as soon as they are identified.

Prevention
Establishment of a steady growth profile. Avoidance of physical stresses on the musculo-skeletal system through careful handling as well as careful design and layout of equipment.

SALMONELLA GALLINARUM, FOWL TYPHOID

Introduction
Disease caused by one of the two poultry-adapted strains of *Salmonella* bacteria, *Salmonella* Gallinarum. This can cause mortality in birds of any age. Broiler parents and brown-shell egg layers are especially susceptible. Chickens are most commonly affected but it also infects turkeys, game birds, guinea fowls, sparrows, parrots, canaries and bullfinches. Infections still occur worldwide in non-commercial poultry but are rare in most commercial systems now. Morbidity is 10–100%; mortality is increased in stressed or immunocompromised flocks and may be up to 100%. The route of infection is oral or via the navel/yolk. Transmission may be transovarian or horizontal by faecal–oral contamination, egg eating etc, even in adults. The bacterium is fairly resistant to normal climate, surviving months, but is susceptible to normal disinfectants.

Signs
Dejection.

Ruffled feathers.
Inappetance.
Thirst.
Yellow diarrhoea.
Reluctance to move.

Post-mortem lesions

Bronzed enlarged liver with small necrotic foci, and/or congestion.
Engorgement of kidneys and spleen.
Anaemia.
Enteritis of anterior small intestine.

Figure 32. Bacterial septicaemia caused by *Salmonella* Gallinarum, in young broiler chickens. The liver on the left is normal, the one on the right is slightly enlarged, pale and shows focal necrosis. Similar lesions can be seen in Pullorum Disease and with other causes of septicaemia in young chicks.

Diagnosis

Isolation and identification. In clinical cases direct plating on Brilliant Green, McConkey and non-selective agar is advisable. Enrichment procedures usually rely on selenite broth followed by plating on selective media. Tube and rapid plate agglutination tests have been the standard serological tests for many years but have only been validated for chickens. LPS-based Elisa assays have been developed but not widely applied commercially. Differentiate from Pasteurellosis, pullorum disease and coli-septicaemia.

Treatment

Amoxycillin, potentiated sulponamide, tetracylines, fluoroquinolones.

Prevention

Biosecurity, clean chicks. As with other salmonellae, recovered birds are resistant to the effects of infection but may remain carriers. Vaccines for fowl typhoid have been used in some areas, both live (usually based on the Houghton 9R strain) and bacterins.

SALMONELLA PULLORUM, PULLORUM DISEASE, 'BACILLARY WHITE DIARRHOEA'

Introduction

Disease caused by one of the two poultry-adapted strains of *Salmonella* bacteria, *Salmonella* Pullorum, this usually only causes mortality in birds up to 3 weeks of age. Occasionally it can cause losses in adult birds, usually brown-shell egg layers. It affects chickens most commonly, but also infects turkeys, game birds, guinea fowls, sparrows, parrots, ring doves, ostriches and peafowl. It still occurs worldwide in non-commercial poultry but is now rare in most commercial systems. Morbidity is 10–80%; mortality is increased in stressed or immunocompromised flocks and may be up to 100%. The route of infection is oral or via the navel/yolk. Transmission may be transovarian or horizontal mainly in young birds and may sometimes be associated with cannibalism. The bacterium is fairly resistant to normal climate, surviving months but is susceptible to normal disinfectants.

Signs

Inappetance.
Depression.

Ruffled feathers.
Closed eyes.
Loud chirping.
White diarrhoea.
Vent pasting.
Gasping.
Lameness.

Post-mortem lesions
Grey nodules in lungs, liver, gizzard wall and heart.
Intestinal or caecal inflammation.
Splenomegaly.
Caecal cores.
Urate crystals in ureters.

Diagnosis
Isolation and identification. In clinical cases direct plating on Brilliant Green, McConkey and non-selective agar is advisable. Enrichment procedures usually rely on selenite broth followed by plating on selective media. Differentiate from Typhoid, Paratyphoid, paracolon, other enterobacteria, chilling and omphalitis

Treatment
Amoxycillin, poteniated sulponamide, tetracylines, fluoroquinolones.

Prevention
Eradication from breeder flocks. As with other salmonellae, recovered birds are resistant to the effects of infection but may remain carriers. Vaccines are not normally used as they interfere with serological testing and elimination of carriers.

SALMONELLOSIS, PARATYPHOID INFECTIONS

Introduction
Salmonellae bacteria other than the species specific sero-types S. Pullorum, S. Gallinarum, and also other than S. Enteritidis and S.

Typhimurium (which are considered separately), are capable of causing enteritis and septicaemia in young birds. Sero-types vary, but *S*. Derby, *S*. Newport, *S*. Montevideo, *S*. Anatum, *S*. Bredeney are among the more common isolates. Even if these infections do not cause clinical disease, their presence may be significant with respect to carcase contamination as a potential source of human food poisoning. They infect chickens, turkeys and ducks worldwide. Morbidity is 0–90% and mortality is usually low. The route of infection is oral and transmission may be vertical as a result of shell contamination. Regardless of the initial source of the infection, it may become established on certain farms, in the environment or in rodent populations. Many species are intestinal carriers and infection is spread by faeces, fomites and feed (especially protein supplements but also poorly stored grain). Certain sero-types are prone to remain resident in particular installations (e.g. *S*. Senftenberg in hatcheries). The bacteria are often persistent in the environment, especially in dry dusty areas, but are susceptible to disinfectants that are suitable for the particular contaminated surfaces and conditions, applied at sufficient concentrations. Temperatures of around 80ºC are effective in eliminating low to moderate infection if applied for 1–2 minutes. This approach is often used in the heat treatment of feed. Predisposing factors include nutritional deficiencies, chilling, inadequate water, other bacterial infections and ornithosis (in ducks).

Signs

Signs are generally mild compared to host-specific salmonellae, or absent.
Dejection.
Ruffled feathers.
Closed eyes.
Diarrhoea.
Vent pasting.
Loss of appetite and thirst.

Post-mortem lesions

In acute disease there may be few lesions.
Dehydration.
Enteritis.
Focal necrotic intestinal lesions.

Foci in liver.
Unabsorbed yolk.
Cheesy cores in caecae.
Pericarditis.

Diagnosis

Isolation and identification. In clinical cases direct plating on Brilliant Green and McConkey agar may be adequate. Enrichment media such as buffered peptone followed by selective broth or semi-solid media (e.g. Rappaport-Vassiliadis) followed by plating on two selective media will greatly increase sensitivity. However this has the potential to reveal the presence of salmonellae that are irrelevant to the clinical problem under investigation. Differentiate from Pullorum/Typhoid, other enterobacteria.

Treatment

Sulphonamides, neomycin, tetracyclines, amoxycillin, fluoroquinolones. Good management. Chemotherapy can prolong carrier status in some circumstances.

Prevention

Uninfected breeders, clean nests, fumigate eggs, all-in/all-out production, good feed, competitive exclusion, care in avoiding damage to natural flora, elimination of resident infections in hatcheries, mills, breeding and grow-out farms. Routine monitoring of breeding flocks, hatcheries and feed mills is required for effective control. Infection results in a strong immune response manifest by progressive reduction in excretion of the organism and reduced disease and excretion on subsequent challenge. Vaccines are not generally used for this group of infections.

SALMONELLOSIS, *S.* ENTERITIDIS AND *S.* TYPHIMURIUM INFECTIONS

Introduction

Salmonella Enteritidis and *S.* Typhimurium are presented separately from other sero-types of Salmonella because, on the one hand, these bacteria are often specifically cited in zoonosis control legislation, and, secondly, because there are differences in the epidemiology as compared to other salmonellae. These are the predominant sero-types associated with human disease in most countries. *Salmonella* Enteritidis, especially phage type 4, has become much more common in both poultry and humans since the early 80s. The prevalence of *S.*Typhimurium has remained relatively stable though the spread of the highly antibiotic-resistant strain DT104 in various farmed species gives some reason for concern. Infections in chickens, turkeys and ducks cause problems worldwide with morbidity of 0–90% and a low to moderate mortality. Many infected birds are culled and others are rejected at slaughter. The route of infection is oral; many species are intestinal carriers and infection may be carried by faeces, fomites and on eggshells. Vertical transmission may be either by shell contamination or internal transovarian contamination of yolk. Feed and feed raw material contamination is less common than for other sero-types. The bacteria are often persistent in the environment, especially in dry dusty areas, but are susceptible to disinfectants that are suitable for the particular contaminated surfaces and conditions, applied at sufficient concentrations. Temperatures of around 80°C are effective in eliminating low to moderate infection if applied for 1–2 minutes. This approach is often used in the heat treatment of feed. Predisposing factors include nutritional deficiencies, chilling, inadequate water and other bacterial infections.

Signs
Dejection.
Ruffled feathers.
Closed eyes.
Diarrhoea.
Vent pasting.
Lost of appetite and thirst.
Stunting in older birds.

Post-mortem lesions
In acute disease there may be few lesions.
Dehydration.
Enteritis.

Focal necrotic intestinal lesions.
Foci in liver.
Unabsorbed yolk.
Cheesy cores in caecae.
Pericarditis.
Perihepatitis.
Misshapen ovules in the ovaries in S.E. infection

Diagnosis

Isolation and identification. In clinical cases direct plating on Brilliant Green and McConkey agar may be adequate. Enrichment media such as buffered peptone followed by selective broth or semi-solid media (e.g. Rappaport-Vassiliadis) followed by plating on two selective media will greatly increase sensitivity. However this has the potential to reveal the presence of salmonellae that are irrelevant to the clinical problem under investigation. Differentiate from Pullorum/Typhoid, other enterobacteria such as *E. coli*. *S.*Enteritidis causes cross-reactions which may be detected with *S.*Pullorum serum agglutination tests. It is possible to detect reactions with specific antigens in agglutination tests but competitive and direct Elisa tests are more commonly used today.

Treatment

Sulphonamides, neomycin, tetracyclines, amoxycillin, fluoroquinolones in accordance with the sensitivity. Good management. Chemotherapy can prolong carrier status in some circumstances.

Prevention

Uninfected breeders, clean nests, fumigate eggs, all-in/all-out production, good feed, competitive exclusion, care in avoiding damage to natural flora, elimination of resident infections in hatcheries, mills, breeding and grow-out farms. Routine monitoring of breeding flocks, hatcheries and feed mills is required for effective control. Early depletion of infected breeding stock is required in some countries such as those of the European Union. Infection results in a strong immune response manifest by progressive reduction in excretion of the organism and reduced disease and excretion on subsequent challenge. Vaccines are increasingly being used for *S.* Enteritidis and *S.* Typhimurium infection; both inactivated (bacterins) and attenuated live organisms.

SALPINGITIS

Introduction

Salpingitis is an inflammation of the oviduct. It is a complex condition of chickens and ducks associated with various infections including *Mycoplasma* and bacteria (especially *E. coli* and occasionally *Salmonella* spp.). Infection may spread downwards from an infected left abdominal air sac, or may proceed upwards from the cloaca. The oviduct is a hollow tube joining the normally sterile environment of the body cavity with the cloaca, which normally has many millions of potentially pathogenic bacteria. The control of infection in this area is probably achieved by ciliated epithelium that mostly wafts a carpet of mucus towards the cloaca. Anything that damages the epithelium or disturbs normal oviduct motility is likely to increase the likelihood of salpingitis. Systemic viral infections that cause ovarian regression or damage to the oviduct or cloaca, are especially prone to increasing salpingitis.

Signs
Sporadic loss of lay.
Death.
Damaged vents, leaking urates.
Distended abdomen.
Some birds may 'lay' a caseous mass of pus (which may be found in a nest or on the egg belt).

Post-mortem lesions
Slight to marked distension of oviduct with exudate.
May form a multi-layered caseous cast in oviduct or be amorphous.
Peritonitis.

Diagnosis
Use the signs to select birds for culling and post-mortem investigation.

Lesions.
Bacteriology of oviduct.

Treatment
Birds with well-developed lesions are unlikely to respond to medication. Use of a suitable antimicrobial may be beneficial for birds in the early stages and if associated with efforts to minimise risk factors.

Prevention
Control any septicaemia earlier in life, use healthy parent flocks, immunise effectively against respiratory viral pathogens common in the area.

SPIKING MORTALITY OF CHICKENS

Introduction
This is a condition characterised by a sudden increase in mortality in young, typically 7–14-day-old, rapidly growing broiler chickens. Birds in good condition die after showing neurological signs. Mortality drops off as sharply as it started. This appears to be a multifactorial condition. Feed intake, and/or carbohydrate absorption are disturbed resulting in a hypoglycaemia. Males are more susceptible than females, probably because they are growing faster. Filtered intestinal contents from affected flocks appear to be capable of reproducing the condition, suggesting a viral component. In order to reproduce the typical condition the affected birds are subject to 4 hours without feed and then a mild physical stress such as spraying with cool water.

Signs
Tremor.
Paralysis.
Coma.
Death.
Orange mucoid droppings.

Post-mortem lesions
Mild enteritis.

Excess fluid in lower small intestine and caecae.
Dehydration.

Diagnosis
Pattern of mortality.
Signs and lesions.

Treatment
Leave affected chicks undisturbed.
Provide multivitamins, electrolytes and glucose solution to flock.
Minimise stress.

Prevention
Good sanitation of the brooding house.
Avoidance of interruptions in feed supply.
Avoidance of physical stress.

SPIROCHAETOSIS

Introduction
The bacterium *Borrelia anserina* infects chickens, turkey, geese, ducks, pheasants, grouse and canaries with morbidity and mortality up to 100%. It is transmitted by arthropods, e.g. *Argas persicus*, and occasionally by infected faeces. The bacterium is poorly resistant outside host but may be carried by *Argas persicus* for 430 days. *Brachyspira pilosicoli*, previously known as *Serpulina pilosicoli*, is an intestinal spirochaete that can be associated with inflammation of the large intestine in a broad range of mammals and birds. It has been associated with typhilitis, diarrhoea, reduced egg production, and egg soiling in chickens.

Signs
Depression.
Cyanosis.
Thirst.
Often diarrhoea with excessive urates.
Weakness and progressive paralysis.
Drops in egg production may be seen in both systemic and intestinal forms

Post-mortem lesions
Marked splenomegaly.
Spleen mottled with ecchymotic haemorrhages.
Liver enlarged with small haemorrhages.
Necrotic foci.
Mucoid enteritis.

Diagnosis
Haematology, isolate in chicken eggs or chicks or poults.

Treatment
Various antibiotics including penicillin.

Prevention
Control vectors, vaccines in some countries.

SPONDYLOLISTHESIS, KINKY-BACK

Introduction
A complex condition of broiler chickens with a morbidity of 1–5% and low mortality associated with rapid growth and possibly having a genetic component.

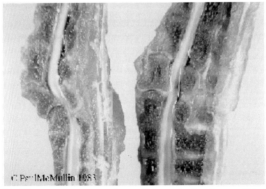

Figure 33. These are cross-sections of the spinal column of 4–5-week-old broilers. The sample on the right is normal. That on the left shows the typical lesion of spondylolisthesis, with rotation of the body of the vertebra and the pinching of the spinal column that causes paralysis.

Signs
Sitting on hock or rumps, legs extended forward.

Use of wings to help in walking.
May walk backwards.

Post-mortem lesions
Anterior–posterior rotation of the bodies of the last or penultimate thoracic vertebrae resulting in scoliosis just anterior to the kidney.

Diagnosis
Symptoms, lesions.

Treatment
None.

Prevention
Selection for satisfactory conformation in primary breeding.

SPRADDLE LEGS OR SPLAY LEG

Introduction
Occurs in all species in the recent hatchling.

Signs
Legs splay and chick is unable to stand.

Post-mortem lesions
Gross lesions are not usually evident.

Diagnosis
Clinical signs.

Treatment
None.

Prevention
All measures that improve chick vitality at hatching are beneficial in reducing this problem. These include good breeder nutrition, avoidance of excessive hatching egg storage, and careful monitoring of incubation conditions. The type of chick box liners and the feeding papers used in the brooding house can also influence the occurrence of the condition.

Staphylococcosis, Staphylococcal arthritis, bumble foot

Introduction
Caused by *Staphylococcus* bacteria, mainly *S. aureus* and seen in chickens and turkeys worldwide. Morbidity is usually low and mortality 0–15% though affected birds will often be culled on humane grounds. Infection is usually by the respiratory route with an incubation period of 2–3 days seen after artificial infection. Wounds, either accidental or induced by interventions such as beak trimming, and toe trimming may be a portal of entry with subsequent spread via the bloodstream to the typical sites of lesions. Damaged skin due to nutritional deficiencies (such as of biotin) may also be a point of entry. Transmission occurs in the hatchery and in the general farm environment, and by fomites. Predisposing factors include reovirus infection, chronic stress, trauma, and imunosuppression.

Signs
Ruffled feathers.
Lameness.
Low mobility.
Swollen above the hock and around the hocks and feet.
Some sudden deaths from acute septicaemia if very heavy challenge.

Post-mortem lesions
Tenosynovitis, most commonly in the plantar area of the foot or just above the hock joint. This may progress to abscess formation in these areas.
Infected joints may have clear exudate with fibrin clots.

Diagnosis
Lesions, isolation and identification of pathogen. Differentiate from septicaemia or tenosynovitis due to Colibacillosis, *Salmonella* spp., *Mycoplasma* spp., especially *M. synoviae*.

Treatment
Antibiotics, in accordance with sensitivity.

Prevention

Good hygiene in the nest, the hatchery and in any intervention or surgery (processing, e.g. toe clipping). Possibly vaccination against reovirus infection, particularly of parent birds. Good management, low stress and prevention of immunosuppression from any cause will all tend to help. Competitive exclusion with a non-pathogenic *Staphylococcus* has been shown to be effective (no commercial products yet available based on this technology). Recovered birds may have some immunity but vaccination with staphylococci has not been found to be helpful in preventing the disease to date.

SUDDEN DEATH SYNDROME, 'FLIPOVER'

Introduction

A condition of broiler chickens of unknown cause, possibly metabolic. It can be induced by lactic acidosis and about 70% of birds affected are males.

Signs

Sudden death in convulsion, most are found lying on their back.

Post-mortem lesions

Intestine filled with feed.
Haemorrhages in muscles and kidneys.
The atria of the heart have blood, the ventricles are empty.
Serum accumulation in lung (may be little if examined shortly after death).
Livers heavier than those of pen-mates (as a percentage of bodyweight.).

Diagnosis

Birds found on back with lack of other pathology.

Treatment

None possible.

Prevention

Lowering carbohydrate intake (change to mash), feed restriction, lighting programmes, low intensity light, use of dawn to dusk simulation and avoidance of disturbance.

TAPEWORMS, CESTODES

Introduction
Cestodes are tapeworms that are seen in many species; they may not be host specific. Most have intermediate invertebrate hosts such as beetles or earthworms.

Signs
It is doubtful if any signs are produced under most circumstances.

Post-mortem lesions
Occupy space in intestine and create small lesions at point of attachment.

Diagnosis
Identification of the presence of the worms at post-mortem examination.

Treatment
Flubendazole is effective at a 60 ppm in diet, however it may not have a zero withdrawal in commercial egg layers. Check your local regulations.

Prevention
Control the intermediate hosts, or birds' access to them.

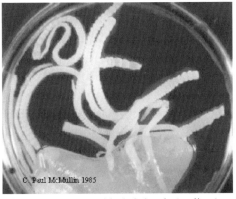

Figure 34. Mature tapeworms with their heads (scolices) embedded in the intestine.

TIBIAL DYSCHONDROPLASIA, TD

Introduction
A complex condition seen in chickens, turkeys and ducks. It may be associated with rapid growth and have a nutritional factor.

Signs
There are usually no signs unless the condition is severe.
Swelling and bowing in the region of the knee joints.
Lameness.

Post-mortem lesions
Plug of cartilage in proximal end of tibia, distal tibia, and proximal metatarsus, in decreasing order of frequency.
Microscopically – a mass of avascular cartilage with transitional chondrocytes, small ovoid lacunae and more matrix than normal.

Diagnosis
Gross pathology; mild lesions may require histology to distinguish from other problems. Lixiscope may identify *in vivo* as early as 2 weeks of age. Differentiate from rickets.

Treatment
None.

Prevention
Genetic selection using the Lixiscope to identify bone phenotype, modifications of calcium and phosphorus ratios, Vitamin D3 supplementation, chloride levels and acid/base balance.

TICKS

Introduction
Argas persicus is an external parasite of poultry and wild birds, and is also found on mammals. It is more common in warm climates, spends little time on host, and may also transmit spirochaetosis and *Pasteurella* infection. These parasites are more likely to be a problem of small scale poultry production in the tropics, than in commercial poultry in temperate climates.

Signs
Anaemia.
Skin blemishes.
Occasionally paralysis from toxins in the tick saliva.
Emaciation.
Weakness.
Reduced productivity.

Post-mortem lesions
Anaemia.

Diagnosis
Identification of the presence of the parasites.

Treatment
Elimination of cracks and crevices in the poultry housing. Insecticide sprays in these areas are more likely to be effective than treatment of the birds.

Prevention
As for red mite control.

TRICHOMONIASIS, CANKER, FROUNCE

Introduction
Trichomonas gallinae is a protozoan parasite of pigeons and doves (canker), raptors (frounce), turkeys and chickens. It has variable pathogenicity. Morbidity is high and mortality varies but may be high. Transmission is via oral secretions in feed and water, and crop milk.

Signs
Mouth open.
Drooling and repeated swallowing movements.
Loss of conditon.
Watery eyes in some birds.
Nervous symptoms (rare).

Post-mortem lesions
Yellow plaques and raised cheesy masses in mouth, pharynx, oesophagus, crop and proventriculus.

Raptors may have liver lesions.

Diagnosis

Lesions, large numbers of protozoa in secretions. Differentiate from Pox and candidiasis.

Treatment

Previously approved products included dimetridazole, nithiazide, and enheptin. Organic arsenical compounds (where approved) and some herbal products may be of some benefit in managing this problem in food animals.

Prevention

Eliminate stagnant water, eliminate known carriers, add no new birds.

TUBERCULOSIS

Introduction

A bacterial infection, caused by *Mycobacterium avium*, of poultry, game birds, cage birds etc. Morbidity and mortality are high. Transmission is via faecal excretion, ingestion, inhalation, offal and fomites. The disease has a slow course through a flock. The bacterium resists heat, cold, water, dryness, pH changes and many disinfectants.

Signs

Severe loss of weight with no loss of appetite.
Pale comb.
Diarrhoea.
Lameness.
Sporadic deaths.

Post-mortem lesions

Emaciation.
Grey to yellow nodules attached to intestine.
Granulomas in liver, spleen and many other tissues, even bone marrow.

Diagnosis
Isolation, acid-fast stain in tissues, TB test. Differentiate from Lymphoid leukosis.

Treatment
None recommended.

Prevention
Market after one season, hygiene, cages, elimination of faeces etc.

TWISTED LEG

Introduction
A complex condition seen in chickens and turkeys. Morbidity is 1–20% and mortality is low. Factors involved may include genetics, nutrition, environment and high growth rate.

Signs
Lameness.
Distortion at hock.
Valgus/varus.
Various angulations of leg.
Gastrocnemius tendon may slip.

Post-mortem lesions
Linear twisting of growth of long bones.
Changed angulation of tibial condyles; intertarsal angulation up to 10% is physiological, greater than 20% is abnormal.

Diagnosis
Symptoms, lesions. Differentiate from perosis, arthritis and synovitis.

Treatment
None.

Prevention
Selection for good conformation in primary breeding. Adoption of optimal nutrition and growth patterns for the economic objectives of the production system.

Ulcerative Enteritis, Quail disease

Introduction
An acute, highly contagious disease of chickens and quail caused by the bacterium *Clostridium colinum* and characterised by ulcers of the intestines and caecae. It can start suddenly and cause high mortality: 100% in quail and 10% in chickens. Turkeys, game birds and pigeons may also be affected. The condition occurs worldwide. The route of infection is oral and transmission is from faeces of sick or carrier birds or via flies. The bacterium resists boiling for 3 minutes. Predisposing factors include Coccidiosis (especially *E. necatrix, E. tenella, and E. brunetti*), IBDV and overcrowding.

Signs
Listlessness.
Retracted neck.
Drooping wings.
Partially closed eyes.
Ruffled feathers.
Diarrhoea.
Anaemia.
Watery white faeces (quail).

Post-mortem lesions
Deep ulcers throughout intestine, but mainly ileum and caecae, which may coalesce and may be round or lenticular.
Pale yellow membranes,.
Peritonitis (if ulcers penetrate).
Blood in intestine.
Necrotic foci in liver.

Diagnosis
A presumptive diagnosis may be made on history and lesions. Confirmation is on absence of other diseases and isolation of *Cl. colinum* in anaerobic conditions (the agent is often present in pure culture in liver). Differentiate from histomonosis ('Blackhead'), necrotic enteritis, coccidiosis, salmonellosis, trichomoniasis.

Treatment
Streptomycin (44 gm/100 litres water), Bacitracin, Tetracyclines, penicillin (50–100 ppm in feed), amoxycillin, multivitamins. Response to treatment should occur in 48 to 96 hours. Treat for coccidiosis if this is a factor.

Prevention
Infection-free birds, all-in/all-out production, low level antibiotics as per treatment, possibly probiotics.

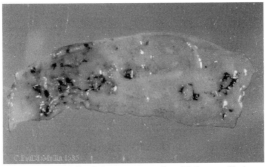

Figure 35. Wall of the caecum of a chicken suffering from ulcerative enteritis

VIBRIONIC HEPATITIS, AVIAN INFECTIOUS HEPATITIS

Introduction
An insidious onset disease of chickens caused by *Vibrio* bacteria. Morbidity is low. Transmission is by faecal contamination, birds remaining carriers for months, and disease is precipitated by stress. The infective agent is rather resistant to environment and disinfectants.

Signs
Dejection.
Diarrhoea.
Loss of condition.
Inappetance.
Pale comb and wattles.
Scaly comb.

Jaundice.
Drop in production/weight gain.

Post-mortem lesions
Focal hepatic necrosis in 10% of affected. Foci often stellate, or there may be a cauliflower-like 'spotty liver'.
Haematocysts under capsule.
Swelling of organs.
Catarrhal enteritis.

Diagnosis
History, lesions, isolation of infective agent from bile. Differentiate from leukosis, histomonosis, ulcerative enteritis, fowl cholera, and typhoid.

Treatment
Erythromycin, fluoroquinolones.

Prevention
Hygiene, depopulate, obtain birds free of disease, contain stressors.

VIRAL ARTHRITIS

Introduction
Viral Arthritis is the classic, but by no means the only, manifestation of reovirus infection of chickens; at least 5 sero-types of virus occur. Morbidity is high but mortality is usually low. Transmission is by faecal contamination, and good both laterally and vertically. Birds remain carriers for over 250 days. The virus is resistant to heat, ether, chloroform, pH and environmental factors. Reoviruses vary markedly in pathogenicity and the tissue damaged. Some can cause other disease syndromes such as early chick mortality and malabsorption syndrome (see page 150). Some strains have shown severe systemic disease including pericarditis in chickens. Others have caused immunosuppression by damaging the cloacal bursa in ducks.

Signs
Lameness.

Low mobility.
Poor growth.
Inflammation at hock.
Swelling of tendon sheaths.
Unthriftiness.
Rupture of gastrocnemius tendons.

Post-mortem lesions
Swelling and inflammation of digital flexor and metatarsal extensor tendon sheaths.
Foot pad swelling.
Articular cartilages may be ulcerated.
Haemorrhage in tissues.
Fibrosis in chronic cases.

Diagnosis
Diagnosis may be based on the history, lesions, IFA and rising antibody titre. Isolation may be readily achieved in CE yolk sac and CAM and also cell cultures (CE kidney or liver cells). Serology may be by DID, FAT or Elisa. 'Silent' infections (not associated with obvious disease) are common. Differentiate from mycoplasmosis, salmonellosis, Marek's, *Pasteurella*, erysipelas.

Treatment
None.

Prevention
Vaccination is ideally carried out by administering a live vaccine in rear followed by an inactivated vaccine prior to coming into lay. Most vaccines are based on strain 11/33. Rear birds in all-in/all-out production systems.

VISCERAL GOUT, NEPHROSIS, BABY CHICK NEPHROPATHY

Introduction
Visceral gout is the deposition of white urates, which are normally excreted as a white cap on well formed faeces, in various tissues.

Urates are also often deposited in joints and in the kidney. This condition can occur as an individual problem at any age. Outbreaks are seen in young chicks in the first week of life (baby chick nephropathy) or in flocks suffering kidney damage, or reduced water intake. All poultry species are susceptible. The kidney damage can arise from infection with certain strains of Infectious Bronchtiis virus, exposure to some mycotoxins or inadequate water intake (often because the birds have not adapted to a new type of drinker). Baby Chick Nephropathy can be due to inappropriate egg storage conditions, excessive water loss during incubation or during chick holding/transport, or inadequate water intake during the first few days of life. Very low humidity in brooding will also increase the likelihood of this problem. The timing of mortality is a reasonable guide as to the source of the problem. In Pekin/mallard ducks the condition is almost always due to inadequate water intake, whereas in muscovy ducks it is seen in breeders allowed to continue laying for over 24 weeks without a rest.

Signs
Depression.
Low feed intake and growth.

Post-mortem lesions
Chalky white deposits on pericardium, liver, air sacs, peritoneum
Similar deposits may be present in joints and are usually present in the kidney.

Diagnosis
Lesions.

Treatment
This is based on correcting any management errors and encouraging water intake. Avoid any intentional or unintentional restriction in water intake. Sodium bicarbonate at 1g/litre water is mildly diuretic, however it could be counter-productive if water intake is in any way restricted.

Prevention
Careful monitoring of the conditions of hatching egg storage

and incubation with a view to achieving a standard egg weight loss profile. Humidification of holding rooms and chick transporters may also be beneficial. Humidifiers in chick brooding areas are being used increasingly, especially where whole house hot air brooders are in use. Ample supplies of drinkers should be available and filled with water at house temperature prior to the arrival of the chicks.

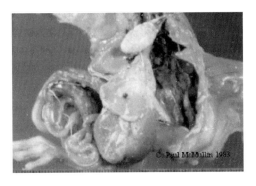

Figure 36. Severe visceral gout in a young chick. There are white chalky deposits around the heart (in the pericardium) on all major abdominal organs, including liver, gizzard and intestines, and even in the tissues of the thigh.

VITAMIN A DEFICIENCY, NUTRITIONAL ROUP

Introduction
Vitamin A deficiency is occasionally seen in chickens and turkeys (insufficient vitamin A during 1–7 weeks of age). As in the case of other nutritional deficiencies, classic signs of deficiency are very rare in commercial poultry fed complete diets.

Signs
Poor growth.
Poor feathering.
Nasal and ocular discharge.

Drowsiness.
Pale comb and wattles.
Eyelids stuck shut with thick exudate.

Post-mortem lesions
Eyelids inflamed and adhered.
Excessive urates in kidneys and ureters.
Pustules in mouth and pharynx.
Microscopically – squamous metaplasia of epithelia.

Diagnosis
Signs, lesions, feed formulation. Differentiate from Infectious coryza, chronic fowl cholera, infectious sinusitis etc.

Treatment
Vitamin A in drinking water.

Prevention
Supplementation of diet with vitamin A, antioxidant, good quality raw materials.

VITAMIN B DEFICIENCIES

Introduction
The B complex vitamins are water soluble and not stored to any significant extent in the body. They act in a broad range of metabolic pathways. Simple deficiency is now rare as diets are usually well supplemented. However, because a continuous supply is required, damage to the intestine or increased demand for some reason may have an effect. Most will reduce productivity, including growth in the young animal, and egg production in the layer. The embryo is particularly dependent on having adequate supplies of vitamins deposited in the egg. Vitamin deficiencies are especially prone to cause problems of hatchability. See the separate discussion under Chondrodystrophy and Fatty Liver and Kidney Syndrome.

Chapter 6 - Diseases and Syndromes: Chickens & Various Species

Signs
These may be summarised:

	Biotin	Cyanocobalmin (B_{12})	Folic Acid	Niacin	Pyridoxine (B_6)	Pantothenic acid	Riboflavin (B_2)	Thiamine (B_1)
Perosis	+-		+-	+-	+-			
Curled Toe Paralysis							+	
Paralysis/'Stargazing'								+
Dermatitis/Scaly Skin						+		
Mouth Lesions				+				
Conjunctivitis								
Anaemia		+	+					
Fatty Liver and Kidney Syndrome	+							
Poor Feathering						+		
Loose feathers	+							
Hatchability problems	+	++	-				++	++
Embryo with clubbed down							+	

Post-mortem lesions
Usually the gross lesions are non-specific.
Some deficiencies induce characteristic microscopic effects.

Diagnosis
Signs, exclusion of specific diseases, response to supplementation. If it is suspected that the vitamin premix may not have been included in the ration (or included at too low a level) it may be appropriate (faster, less expensive) to analyse feed for a marker substance such as manganese rather than testing for vitamin levels.

Treatment
If a specific vitamin deficiency is suspected, drinking water supplementation with that vitamin is ideal and usually results in a rapid response in birds that are still drinking. Good quality multivitamin solutions are beneficial in the supportive care of a range of problems characterised by reduced feed intake. The balance of vitamins present should be similar to the daily nutritional requirement of the stock concerned.

Prevention
Adequate supplementation of the feed with all required vitamins in levels which both support normal productivity but also have enough overage to deal with the increased demands that often occur during periods of disease challenge.

VITAMIN E DEFICIENCY, ENCEPHALOMALACIA, EXUDATIVE DIATHESIS, MUSCULAR DYSTROPHY

Introduction
A spectrum of diseases of chickens and turkeys, occasionally ducklings and other birds, seen worldwide, characterised by oxidation of various tissues and caused by Vitamin E deficiency. The problem is associated with feed rancidity typically in diets with high fat. Encephalomalacia and exudative diathesis tends to be seen in young birds of 1–5 weeks of age. Muscular dystrophy is seen more frequently in older and mature birds.

Signs
- Imbalance.
- Staggering.
- Uncontrolled movements.
- Falling over.
- Paralysis.
- Ventral oedema.
- Green wings.

Post-mortem lesions
- Swollen cerebellum with areas of congestion.
- Haemorrhage.
- Necrosis.
- Blood-stained or greenish subcutaneous oedema.
- Steatitis.
- White streaks in muscle.

Diagnosis
Signs, lesions, feed rancidity, histopathology, response to medication. Differentiate from Encephalomyelitis, toxicities, necrotic dermatitis.

Treatment
Vitamin E and/or selenium in feed and/or water. Broad-spectrum antibiotics where there are extensive skin lesions.

Prevention
Proper levels of vitamin E, selenium, antioxidant, good quality raw materials.

YOLK SAC INFECTION, OMPHALITIS

Introduction
A condition seen worldwide in chickens, turkeys and ducks due to bacterial infection of the navel and yolk sac of newly hatched chicks as a result of contamination before healing of the navel. Disease occurs after an incubation period of 1–3 days. Various bacteria may be involved, especially *E .coli, Staphylococci, Proteus, Pseudomonas*. Morbidity is 1–10% and mortality is high in affected chicks. It is seen where there is poor breeder farm nest hygiene, use of floor eggs, inadequate hatchery hygiene or poor incubation conditions, for example poor hygiene of hatching eggs, 'bangers', and poor hygiene of setters, hatchers or chick boxes. Inadequate incubation conditions resulting in excessive water retention and slowly-healing navels and 'tags' of yolk at the navel on hatching also contribute to the problem.

Signs
Dejection.
Closed eyes.
Loss of appetite.
Diarrhoea.
Vent pasting.
Swollen abdomen.

Post-mortem lesions
Enlarged yolk sac with congestion.
Abnormal yolk sac contents (colour, consistency) that vary according to the bacteria involved.

Diagnosis
A presumptive diagnosis is based on the age and typical lesions.

Confirmation is by isolation and identification of the bacteria involved in the internal lesions. Differentiate from incubation problems resulting in weak chicks.

Treatment

Antibiotics in accordance with sensitivity may be beneficial in the acute stages, however the prognosis for chicks showing obvious signs is poor; most will die before 7 days of age.

Prevention

Prevention is based on a good programme of hygiene and sanitation from the nest through to the chick box (e.g. clean nests, frequent collection, sanitation of eggs, exclusion of severely soiled eggs, separate incubation of floor eggs etc. There should be routine sanitation monitoring of the hatchery. Multivitamins in the first few days may generally boost ability to fight off mild infections.

DISEASES AND SYNDROMES: PREDOMINANTLY OF TURKEYS

Arizona Infection, Arizonosis, Paracolon Infection, *Salmonella* Arizonae

Introduction
Caused by the bacterium *Arizona hinshawii*, renamed *Salmonella* Arizonae. It affects turkeys, mainly in North America, and is not present in the UK turkey population. Mortality is 10–50% in young birds, older birds are asymptomatic carriers. Transmission is vertical, transovarian, and also horizontal, through faecal contamination of environment, feed etc, from long-term intestinal carriers, rodents, reptiles.

Signs
Dejection.
Inappetance.
Diarrhoea.
Vent-pasting.
Nervous signs.
Paralysis.
Blindness, cloudiness in eye.
Huddling near heat.

Post-mortem lesions
Enlarged mottled liver.
Unabsorbed yolk sac.
Congestion of duodenum.
Cheesy plugs in intestine or caecum.
Foci in lungs.
Salpingitis.
Ophthalmitis.
Pericarditis.
Perihepatitis.

Diagnosis
Isolation and identification, methods as per *Salmonella* spp.
Differentiate from salmonellosis, coli-septicaemia

Treatment
Injection of streptomycin, spectinomycin, or gentamycin at the

hatchery is used in some countries. Formerly in-feed medication with nitrofurans was also used.

Prevention
Eradicate from breeder population, fumigation of hatching eggs, good nest and hatchery hygiene, inject eggs or poults with antibiotics, monitor sensitivity.

CHLAMYDIOSIS, PSITTACOSIS, ORNITHOSIS

Introduction
An infection of turkeys, ducks, psittacines, pigeons, man, rarely chickens, caused by *Chlamydia psittaci*, a bacterium of highly variable pathogenicity. It is a 'Scheduled Disease' rarely diagnosed in UK, but occurring probably worldwide. Morbidity is 50–80%, mortality 5–40%. It is transmitted by contact, faecal dust and wild bird carriers, especially pigeons and robins. Egg transmission does not occur. Elementary bodies are highly resistant and can survive in dried faeces for many months. Iodophores and formaldehyde are effective disinfecting agents, phenolics are less so. Intercurrent salmonellosis and, perhaps, other infections may be predisposing factors.

Signs
Respiratory signs.
Greenish-yellow diarrhoea.
Depression.
Weakness.
Inappetance.
Weight loss.
Nasal discharge.
Conjunctivitis.
Occasional transient ataxia in pigeons.
Production drops in naive laying flocks

Post-mortem lesions
Vascular congestion.
Wasting.
Fibrinous pericarditis.

Airsacculitis.
Perihepatitis.
Spleen enlarged and congested, may rupture in pigeons.
Necrotic foci in liver.
Fibrinous pneumonia.
Congested lungs and air sacs in the turkey.

Diagnosis
History, signs, lesions. Intracytoplasmic inclusions are helpful but confirmation requires demonstration of causal organisms (Giemsa stain, IFA). Serology: complement fixation, Elisa and gel diffusion. Differentiate from Duck viral hepatitis, Duck septicaemia.

Treatment
Tetracycline (200–800 ppm in feed for 3–4 weeks) and/or quinolone medication and supervised slaughter.

Prevention
Biosecurity, exclusion of wild birds. Live and inactivated vaccines are protective although the former result in carriers and the latter require several applications.

COCCIDIOSIS OF TURKEYS

Introduction
Infection of turkeys with *Eimeria* spp. This disease is not very common in commercially reared turkeys though most turkey growers receive preventative medication for at least part of their lives. Five species of *Eimeria* have been identified that cause lesions in turkeys, of which two are associated with significant disease effects. *E. meleagrimitis* affects the upper small intestine, while E. *adenoides* affects the caecae and rectum. *E. gallopavonis* and *E. meleagridis* affect the lower small intestine rectum and caecae, while *E. dispersa* is found in the small intestine.

Signs
Huddling.
Weight loss.
Depression.

Watery diarrhoea that may occasionally be blood stained or contain clumps of mucus or shed mucosa.
Tucked appearance, ruffled feathers.

Post-mortem lesions

The affected area of intestine shows thickening of the wall and dilation. The contents may be haemorrhagic or be watery with white material shed from the mucosa.

Diagnosis

Signs, lesions, microscopic exam of scrapings (oocysts, gamonts). Differentiate from necrotic enteritis.

Treatment

Toltrazuril, Sulphonamides (e.g. Sulphaquinoxaline), Amprolium.

Prevention

The ionophore coccidiostats lasalocid and monensin are routinely used in turkey growers, typically to 12 weeks of age. Diclazuril is also used for this purpose. Dosage levels of ionophores may be critical to efficacy and safety. Exposure of previously unmedicated birds to these compounds can cause toxicity. Salinomycin is toxic for turkeys even at very low doses. Avoid use of tiamulin in ionophore treated birds.

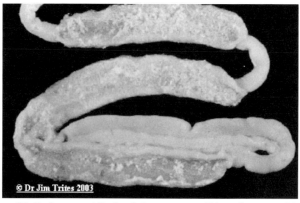

Figure 37. Turkey coccidiosis of the upper small intestine caused by *E. meleagrimitis*. The intestines are dilated, show some spotty congestion and have abnormal contents due to the sloughed epithelium.

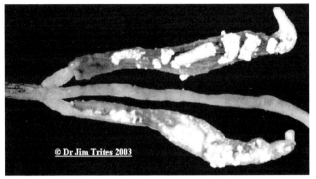

Figure 38. Turkey caecal coccidiosis caused by *E. adenoides*. The exudate can range from semi-liquid to solid white cores.

DISSECTING ANEURYSM, AORTIC RUPTURE

Introduction

A complex, genetic condition of turkeys linked to male sex and high growth rate. It has been suggested that degenerative changes in the wall of the artery and copper deficiency may be factors. A sudden noise or other cause of excitement can lead to an 'outbreak', presumably due to a sudden increase in blood pressure.

Signs

Sudden death with no warning signs, birds found on breast or side.
Skin pale.
Possibly blood in the mouth.

Post-mortem lesions

Carcase anaemic.
Abdominal cavity full of blood.
Haemorrhages in lungs, kidneys, leg muscles, pericardial sac.
Rupture of major blood vessel at base of heart or by the kidneys.
A longtitudinal split of the abdominal aorta is the most common lesion.

Diagnosis
History and lesions.

Treatment
None currently licensed. Reserpine, a tranquilizer, was included in feed at 1 ppm for 3–5 days to reduce blood pressure. Aspirin at 250 ppm in feed or water may be of benefit.

Prevention
Limit feed in birds of 16+ weeks. Reserpine in diet of birds of 4+ weeks (no longer licensed in the UK).

HAEMORRHAGIC ENTERITIS

Introduction
A viral disease of turkeys, similar to pheasant Marble Spleen disease, caused by Type II Adenovirus distinct from classical fowl adenovirus and occurring in most turkey-producing areas. The virulence of the virus varies but morbidity may be 100% and mortality 10–60%. The source of virus is unknown but there is easy lateral spread within flocks, the virus surviving in frozen faeces for months, and weeks in litter. The route of infection is usually oral.

Signs
Sudden deaths.
Blood from vent of moribund birds.
Drop in feed and water consumption.
Diarrhoea.
Course 10–21 days.
May induce immunosupression and precipitate respiratory disease or coccidiosis.

Post-mortem lesions
Petechiae in various tissues.
Intestine distended with blood.
Spleen mottled and enlarged.
Microscopic – spleen shows lymphoid hyperplasia, reticulo-endothelial hyperplasia and intranuclear inclusions.

Chapter 7 - Diseases and Syndromes: Predominantly of Turkeys

Diagnosis

Typical lesions, reproduction with filtered contents, double-immuno-diffusion of splenic extract. Serology: DID, Elisa – sero-conversion frequently occurs in absence of clinical disease.

Treatment

Warmth and good management, convalescent serum, oral tetraycline. Disinfect and 3–4 weeks house rest.

Prevention

All-in/all-out production, good hygiene and biosecurity, good management. Live vaccines are commonly used in many countries at 4–5 weeks of age. Some are produced in live turkeys, some in turkey B-lymphoblastoid cell lines. Maternal antibody may interfere with vaccination. Immunity: there is an early age resistance irrespective of maternal antibody status. Pathogenic strains can depress both B- and T-line lymphocytes for up to 5 weeks following exposure. Immunity to the disease is long lasting.

Figure 39. Turkeys dying with haemorrhagic enteritis commonly have the small intestine distended with blood. In this case a loop of intestine has been opened to show the blood.

HEXAMITIASIS

Introduction

Hexamita meleagridis (pigeons *H. columbae*) is a protozoan parasite of turkeys, pheasants, pigeons, and some game birds. It is transmitted by faeces, fomites, carriers. Inter-species transmission may occur. In commercial ducks a related parasite *Tetratrichomonas* can cause poor growth and drops in egg production.

Signs
Initially birds nervous, chirping.
Later depression.
Inappetance.
Loss in weight.
Frothy, watery diarrhoea.
Terminal coma and convulsions.

Post-mortem lesions
Dehydration.
Intestine flabby with some bulbous dilation, contains excessive mucus and gas.
First half of intestine inflamed.
Caecal tonsils congested.

Diagnosis
Lesions, scrapings from fresh material. Differentiate from transmissible enteritis, paratyphoid, trichomoniasis, histomonosis.

Treatment
Tetracycline, dimetridazole, and also, if possible, increase ambient temperature. Furazolidone, dimetridazole and ipronidazole have been used in the past. The effect of antibiotic may be related to the control of secondary bacterial enteritis.

Prevention
Depopulation, hygiene, all-in/all-out production, avoid inter-species mixing, and mixing groups of different ages.

HISTOMONOSIS, HISTOMONIASIS, BLACKHEAD

Introduction
Histomonas melagridis is a protozoan parasite of turkeys, and occasionally chickens, pheasants and game birds that acts together

with facultative bacteria to produce the condition of Blackhead. This condition has high morbidity and mortality in turkeys. Although chickens are relatively resistant to the condition, significant disease has been seen in breeding chickens and free-range layers. The parasite is ingested in the ova of *Heterakis* worms or as larvae in earthworms or faeces and there is an incubation period of 15–20 days. Outwith earth worms or *H. gallinae* the parasite is easily destroyed. The problem is seen in high-biosecurity facilities, presumably introduced with worm eggs. Within a turkey shed transmission is rapid in spite of the fact that it is difficult to infect birds orally with unprotected parasites. It has recently been demonstrated that infection occurs readily via the cloaca when birds are on contaminated litter.

Signs
Depression.
Inappetance.
Poor growth.
Sulphur-yellow diarrhoea.
Cyanosis of head.
Blood in faeces (chickens).
Progressive depression and emaciation.

Post-mortem lesions
Enlargement of caeca.
Ulcers, caseous cores with yellow, grey or green areas.
Liver may have irregular-round depressed lesions, usually grey in colour, however they may not be present in the early stages, especially in chickens.

Diagnosis
Lesions, scrapings from fresh material.

Treatment
Historically nitro-imidazoles (e.g. dimetridazole), nitrofurans (e.g. furazolidone, nifursol) and arsenicals (e.g. nitarsone) have been used to treat this important disease of poultry. At the time of writing no products of these groups are approved for use in the European Union, and only nitarsone is approved in the USA. Arsenicals are less effective in treatment than they are in

prevention. Some herbal products based on the essential oils (e.g. 'Herban)' have been used with some apparent success though controlled trials and formal approval for this purpose are not recorded. Intensive relittering may help reduce the level of infection, given recent new knowledge on the mechanism of transmission.

Prevention

Good sanitation, avoid mixing species, concrete floors.

Use of an anti-histomonas product in feed where such products are approved but due care with respect to residue avoidance would be required. Regular worming to help control the intermediate hosts. Having both chickens and turkeys on the same property is likely to increase the risk of this disease in turkeys.

LEUCOCYTOZOONOSIS

Introduction

Caused by *Leucocytozoon* species, protozoan parasites of turkeys, ducks, guinea fowl, rarely chickens. The disease has a short course and morbidity and mortality are high. Simulid flies or culicoid midges are intermediate hosts. Survivors may carry infection. Different species of this parasite have been reported from North America, Asia and Africa.

Signs

Sudden onset of depression.
Anorexia.
Thirst.
Loss of equilibrium.
Rapid breathing.
Anaemia.

Post-mortem lesions

Splenomegaly.
Hepatomegaly.
Anaemia.

Microscopically – megaschizonts in brains of birds with nervous signs, schizonts in liver, gametes in erythrocytes.

Diagnosis
Lesions, histology and blood smears.

Treatment
None known.

Prevention
Separation from intermediate hosts, disposal of recovered birds.

LYMPHOPROLIFERATIVE DISEASE (LPD)

Introduction
A Type C retrovirus infection of turkeys with low morbidity and high mortality.

Signs
Depression.
Emaciation.
Loss of weight.
Persistent low mortality.
Enlargement of abdomen, liver or bursa.
May be asymptomatic.

Post-mortem lesions
Focal tumours, especially in enlarged spleen.

Diagnosis
History, age, lesions, cytology. Differentiate from Reticuloendotheliosis.

Treatment
None.

Prevention
Good hygiene, all-in/all-out production, control arthropods, in future eradication.

Mycoplasma gallisepticum infection, M.G., Infectious sinusitis – Turkeys

Introduction

A slow onset chronic respiratory disease of turkeys often with severe sinusitis and associated with *Mycoplasma gallisepticum* infection. It is seen worldwide, though in many countries this infection is now rare in commercial poultry. Morbidity is low to moderate and mortality low. The route of infection is via the conjunctiva or upper respiratory tract with an incubation period of 6–10 days. Transmission may be transovarian, or by direct contact with birds, exudates, aerosols, and fomites. Recovered birds remain infected for life; subsequent stress may cause recurrence of disease. The infectious agent survives for only a matter of days outwith birds, although prolonged survival has been reported in egg yolk and allantoic fluid, and in lyophilised material. Survival seems to be improved on hair and feathers. Stress, malnutrition, intercurrent viral disease such as Newcastle disease (even lentogenic) and Turkey Rhinotracheitis are predisposing factors.

Signs
Coughing.
Nasal and ocular discharge.
Swollen sinuses.
Slow growth.
Leg problems.
Stunting.
Inappetance.

Post-mortem lesions
Swollen infraorbital sinuses, often with inspissated pus.
Airsacculitis.
Pericarditis.
Perihepatitis.

Diagnosis
Lesions, serology, isolation and identification of organism, demonstration of specific DNA (commercial kit available). Culture, of swabs taken from the trachea or lesions, requires

inoculation in mycoplasma-free embryos or, more commonly in Mycoplasma Broth followed by plating out on Mycoplasma Agar. Suspect colonies may be identified by immunofluorescence. Serology: serum agglutination is the standard screening test, suspect reactions are examined further by heat inactivation and/or dilution. HI may also be used. Suspect flocks should be re-sampled after 2–3 weeks. Some inactivated vaccines induce 'false positives' in serological testing. PCR is possible if it is urgent to determine the flock status. Differentiate from viral respiratory disease, especially Turkey Rhinotracheitis.

Treatment

Tilmicosin, tylosin, spiramycin, tetracyclines, fluoroquinolones. Effort should be made to reduce dust and secondary infections.

Prevention

Eradication of this infection has been the central objective of official poultry health programmes in most countries. These are based on purchase of uninfected poults, all-in/all-out production, and biosecurity. In some circumstances preventative medication of known infected flocks may be of benefit.

MYCOPLASMA IOWAE INFECTION, M.I.

Introduction

Infection with *Mycoplasma iowae* is seen in turkeys, and can occur in other poultry and wild bird species. Infection occurs via the conjunctiva or upper respiratory tract and transmission among poults may be vertical, venereal or horizontal.

Signs

Embryonic mortality.
Reduced hatchability.

Post-mortem lesions

Stunted embryos with hepatitis and enlarged spleens, some with down abnormality. Leg lesions can be shown in inoculated birds but do not seem to occur naturally, perhaps because all affected embryos fail to hatch.

Diagnosis
Isolation and identification of the causative organism from dead-in-shell embryos.

Treatment
Pressure differential dipping has been used where breeder flocks are infected. Specialised dip tanks are subjected to negative pressure which partially collapses the air cell of the submerged eggs. When the vacuum is released the eggs draw in antibiotic solution which is subsequently absorbed through the shell membranes. This can increase hatchability by 5–10% in this situation. Antibiotics that have been used are tylosin and enrofloxacin.

Prevention
Eradication of the infection from breeding stock, purchase of M.i.-free stock, maintenance of this status by good biosecurity. For eradication purposes egg injection provides more consistent dosing per egg than does egg dipping.

MYCOPLASMA MELEAGRIDIS INFECTION, M.M.

Introduction
A disease of turkeys characterised by respiratory and skeletal problems caused by *Mycoplasma meleagridis*. The organism has also been isolated from raptors, it occurs in most turkey-producing countries but is now much rarer in commercial stock. In adult birds though infection rates are high, morbidity may be minimal. Pathogenicity is quite variable. Mortality is low, though up to 25% of infected birds show lesions at slaughter. Infection is via the conjunctiva or upper respiratory tract with an incubation period of 6–10 days. Transmission is venereal in breeders, with transovarian and then lateral spread in meat animals. Infected eggs result in widespread distribution of infection and increased risk of further vertical transmission. The infective agent does not survive well outside the bird. Predisposing factors include stress and viral respiratory infections.

Signs
Reduced hatchability.
Slow growth.
Leg problems.
Stunting.
Mild respiratory problems.
Crooked necks.
Infected parents may be asymptomatic.

Post-mortem lesions
Airsacculitis in infected pipped embryos and poults.
Airsacculitis (rarely) seen in adult birds.

Diagnosis
Lesions, serology, isolation and identification of organism, demonstration of specific DNA (commercial kit available). Culture requires inoculation in mycoplasma-free embryos or, more commonly in Mycoplasma Broth followed by plating out on Mycoplasma Agar. Suspect colonies may be identified by immuno-fluorescence. Serology: SAG used routinely – culture used to confirm. Differentiate from *Mycoplasma gallisepticum*, *Mycoplasma synoviae*, other respiratory viruses.

Treatment
Tylosin, spiramycin, tetracyclines, fluoroquinolones. Effort should be made to reduce dust and secondary infections.

Prevention
Eradication of this infection is also possible using similar techniques as for *Mycoplasma gallisepticum*. These are based on purchase of uninfected poults, all-in/all-out production, and biosecurity. Infected males are particularly prone to transmit infection and may warrant special attention. In some circumstances preventative medication of known infected flocks may be of benefit. Vaccines are not normally used. Infected birds do develop some immunity. Birds infected from their parents seem to be immuno-tolerant and particularly prone themselves to transmit.

Osteomyelitis complex, Turkey

Introduction
An association has been shown between green discolouration of the livers of turkey growers at slaughter and the occurrence of inflammatory lesions of bones and/or joints. The green discolouration is used to identify carcasses that require closer inspection of the other tissues.

Signs
None.

Post-mortem lesions
Green discolouration of the liver at slaughter.

Diagnosis
Lesions. Gram-variable pleomorphic bacteria have been isolated from affected livers and bones.

Treatment
Not applicable – only identified at slaughter.

Prevention
Unknown.

Paramyxovirus-3

Introduction
Paramyxovirus PMV-3 infects turkeys, occasionally chickens and psittacine cage birds. The infection is transmitted by birds, including wild bird contact and fomites. Vertical transmission does not usually occur. Mortality is usually low in poultry.

Signs
Depression.
Inappetance.
Coughing.
Drop in egg production (turkeys).
Loss of shell pigment

Nervous symptoms (psittacines).
High mortality (cage birds).

Post-mortem lesions
No specific lesions.

Diagnosis
Isolation in CE, HA+, HI with ND serum or DID (less cross reactions), IFA.

Treatment
No specific treatment, antibiotics to control secondary bacteria.

Prevention
Quarantine, biosecurity, all-in/all-out production. Vaccination has been used in turkeys but may not be cost-effective.

PARAMYXOVIRUS-6

Introduction
Paramyxovirus PMV-6 infection of turkeys. The infection is transmitted by birds, including wild bird contact and fomites. The infection is inapparent in ducks and geese. Vertical transmission does not usually occur. Mortality is 0–5%.

Signs
Reduction in egg production.
Mild respiratory signs.

Post-mortem lesions
No specific lesions.

Diagnosis
Isolation in CE, HA+, HI with ND serum or DID (less cross reactions), IFA.

Treatment
No specific treatment, antibiotics to control secondary bacteria.

Prevention
Quarantine, biosecurity, all-in/all-out production.

PEMS AND SPIKING MORTALITY OF TURKEYS

Introduction
Poult Enteritis and Mortality Syndomre (PEMS) was first identified in high density turkey producing areas of the South Eastern USA in 1991. It is an infectious and transmissible cause of sudden increases in mortality in turkeys between 7 and 28 days of age. A less acute form of the disease appears to produce more of a lingering mortality. In both cases mortality can be high and can be associated with a marked depression in growth. A range of viruses have been isolated from affected flocks. To replicate the condition in full it appears to be necessary to include bacteria in the inoculum. This syndrome is clearly distinguished from typical viral enteritis in young turkeys because of the high mortality and severe growth depression.

Signs
Initial hyperactivity and increased vocalisation.
Increased water consumption.
Picking at feed, eating litter.
Reduced feed consumption and growth.
Increasing weakness, huddling, seeking heat.
Droppings watery, pale brown.
Wet litter.
Weight loss.

Post-mortem lesions
Dehydration.
Muscle atrophy.
Emaciation.
Liquid intestinal contents.
Caseous cores in bursae (late in the process).

Diagnosis
Signs, lesions.

Treatment
A range of antibiotics and arsenicals have been used to control the bacterial component of this condition. Fluoroquinolone antimicrobials appear to be especially effective. A highly

digestible diet with fat at 7.5–8% will encourage feed intake and recovery, multivitamins and milk replacer products are helpful for nutritional support in the acute phase.

Prevention
Biosecurity to reduce the introduction of new infections into brooding facilities. All-in/all-out production. Effective terminal disinfection. Good quality diets of a suitable form – mash and poor quality crumbles increase risk.

SHAKY LEG SYNDROME

Introduction
A condition of turkeys with a course of often 3–4 weeks, possibly associated with tendon injury. Morbidity is 10–20%.

Signs
Stand on toes shaking.
Shaking or quivering of legs after rising.

Post-mortem lesions
None.

Diagnosis
Clinical examination, exclusion of other problems.

Treatment
Multivitamins, aspirin may be helpful if locally approved.

Prevention
Care in transport of brooded poults, releasing poults into larger areas and in handling heavier birds.

TRANSMISSIBLE ENTERITIS, BLUECOMB

Introduction
Some coronaviruses involved in this condition are considered to be bovine coronaviruses (or closely related thereto); others are avian coronaviruses closely related to infectious bronchitis virus. The infection is seen in turkeys and sometimes pheasants. Morbidity is

close to 100% and mortality is 5–100%. Transmission is lateral with birds and faeces, with rapid spread. The course of the disease is about 2 weeks and birds then remain carriers for months. The viruses survive in frozen faeces for months.

Signs
Anorexia.
Depression.
Frothy diarrhoea.
Subnormal body temperature.
Darkening of the head.
Loss of weight.
Huddling.

Post-mortem lesions
Dehydration.
Emaciation.
Enteritis with petechial haemorrhages.
Froth ingesta.
Sometimes casts.
Pancreas chalky.
Urates in kidneys/ureters.

Diagnosis
Reproduction with filtered contents, isolation in turkey eggs, electron microscopy, immuno-fluorescence, Elisa. Commercial M4I IB Elisa will detect some infections.

Treatment
Warmth and good management, antibiotics, milk replacer, potassium chloride. Disinfect and rest houses for 3–4 weeks.

Prevention
All-in/all-out production, good hygiene and biosecurity, good management.

TURKEY CORYZA

Introduction
A condition of turkeys possibly associated with *Bordetella avium*

(previously designated *Alkaligenes faecalis*) infections. This bacterium usually causes mild respiratory disease in only a proportion of the flocks, and low mortality. It can be synergistic with other respiratory pathogens such as *Ornithobacterium*, *Pasteurella* spp, *Mycoplasma* spp. and pneumoviruses.

Signs
Decreased appetite, weight gain and feed efficiency.
Loss of voice.
Ocular and nasal discharge.
Conjunctivitis.
Snick.
Dyspnoea.

Post-mortem lesions
Excessive mucus in nasal passages and trachea.
If there is secondary *E. coli* infection then pneumonia, airsacculitis and perihepatitis.

Diagnosis
Clinical signs, isolation of agent.

Treatment
Antibiotics are not very effective, control respiratory stressors, chlorination of drinking water.

Prevention
All-in/all-out production. Good drinking water hygiene.

TURKEY RHINOTRACHEITIS (IN REAR)

Introduction
A disease of turkeys caused by the viruses of the Pneumovirus genus, Paramyxoviridae. Morbidity is 10–100% and mortality 1–30%. Rapid transmission occurs laterally, possibly involving fomites; vertical transmission is uncertain, maternal antibody may protect.

Signs
Decreased appetite, weight gain and feed efficiency.
Loss of voice.

Ocular and nasal discharge.
Conjunctivitis.
Snick.
Dyspnoea.
Sinusitis.

Post-mortem lesions
Serous rhinitis and tracheitis, sometimes pus in bronchi.
If there is secondary *E. coli* infection then pneumonia, airsacculitis and perihepatitis.

Diagnosis
Clinical signs, serology (using an Elisa test to demonstrate rising titre), isolation and identification of the ciliostatic virus.

Treatment
Antibiotics are not very effective, control respiratory stressors, chlorinate drinking water.

Prevention
All-in/all-out production. Vaccination at day-old seems to be most effective.

TURKEY RHINOTRACHEITIS (ADULT)

Introduction
A disease of turkeys caused by the viruses of the Pneumovirus genus, Paramyxoviridae. Morbidity is 10–100% and mortality 1–30%. Rapid transmission occurs laterally, possibly involving fomites; vertical transmission is uncertain, maternal antibody may protect.

Signs
Decreased appetite.
Loss of voice.
Ocular and nasal discharge.
Sudden drop in production that lasts 2–3 weeks.
Eggs depigmented and thin-shelled.

Post-mortem lesions
Serous rhinitis and tracheitis.

Diagnosis
Signs.
Serology, isolation of ciliostatic agent.

Treatment
Antibiotics are not very effective, control respiratory stressors, chlorinate drinking water.

Prevention
All-in/all-out production, vaccination – live primer followed by inactivated.

TURKEY VIRAL HEPATITIS

Introduction
An infection of turkeys caused by an unidentified virus. Morbidity varies, mortality is 1–25%. Transmission may be by direct or indirect contact and possibly vertical. Birds remain carriers for up to 16 days.

Signs
Signs are usually subclinical.
Depression and sporadic deaths in birds in good condition.

Post-mortem lesions
Grey foci (c. 1 mm) coalescing.
Congestion.
Focal haemorrhage.
Bile staining.
Grey to pink foci in the pancreas.
Microscopically – no inclusion bodies.

Diagnosis
Isolation in CE, lesions pathognomonic. Differentiate from histomonosis, bacterial and fungal infections.

Treatment
None; with good management many birds recover.

Prevention
Good sanitation and management.

DISEASES AND SYNDROMES: PREDOMINANTLY OF DUCKS AND GEESE

Anatipestifer Disease, New Duck Syndrome, Duck Septicaemia

Introduction
An acute or chronic septicaemic disease caused by *Riemerella anatipestifer*, syn *Pasteurella*, or *Moraxella* a. It affects ducks of any age, sometimes turkeys, and may also be isolated from chickens, game birds and wild waterfowl. Mortality is 2–75% in young ducks. Transmission is mainly direct, bird-to-bird, via toenail scratches, especially of the duckling foot, or through respiratory epithelium during respiratory disease. It can also be by faecal contamination of feed, water or the environment where survival of the infectious agent may be prolonged. Adverse environmental conditions and pre-existing disease are predisposing factors.

Signs
Weakness.
Neck tucked in.
Head/neck tremor.
Ataxia.
Disinclined to walk.
Incoordination.
Dyspnoea.
Ocular and/or nasal discharge.
Hyperexcitability

Post-mortem lesions
Perihepatitis without much smell or liver damage.
Pericarditis.
Airsacculitis.
Enlarged liver and spleen.
Occasionally fibrinous meningitis.
Salpingitis
Purulent synovitis.
Chronic arthritis, sometimes with erosions of the joint cartilage.

Diagnosis
Lesions, isolation and identification of organism – blood or chocolate agar in candle jar or 5% CO_2. Differentiate from duck

viral enteritis, duck viral hepatitis, fowl cholera, colibacillosis, coccidiosis, chlamydiosis.

Treatment

Sulphonamides and potentiated sulphonamides are the products most commonly recommended for drinking water application. Subcutaneous injections of penicillin + dihydrostreptomycin, or streptomycin + dihydrostreptomycin are also highly effective.

Prevention

Good husbandry and hygiene, rigid depopulation and disinfection, adequate protection, 'hardening off', correct house relative humidity, sulphonamides in feed. Inactivated and attenuated vaccines available in some countries. Autogenous bacterins sometimes used.

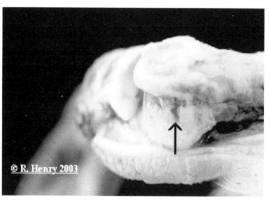

Figure 40. Erosions of the cartilage of the hock joint in a duck with chronic *Riemerella anatipestifer* infection

COCCIDIOSIS, KIDNEY

Introduction

A disease of geese caused by *Eimeria truncata* that can cause high mortality in geese of 3–12 weeks of age, it can also infect Barbary ducks and swans.

Signs

Depression.

Weakness.
Diarrhoea – faeces tend to be whitish.
Reduced feed intake.

Post-mortem lesions
Enlarged kidneys.
Kidneys light grey to greyish pink.
Tiny white foci and petechiae in the kidneys.

Diagnosis
Lesions, presence of coccidial stages in fresh scrapings of kidney lesions.

Treatment
Controlled trials of treatments have not been published.

Prevention
Hygiene.

COCCIDIOSIS, INTESTINAL, OF DUCKS AND GEESE

Introduction
Young ducks and geese may suffer from *Eimeria* spp infection. In the goose *E. anseris* is the most important, while in ducks *Tyzzeria perniciosa* is most pathogenic. *Tyzerria* has eight sporocysts in each oocyst, compared to four per oocyst for *Eimeria*. Coccidiosis occurs only very rarely in commercially reared ducks in the UK.

Signs
Sudden death.
Depression.
Blood-stained vent.
Tucked appearance.

Post-mortem lesions
Massive haemorrhage in upper small intestine.

Diagnosis
Signs, lesions, microscopic examination of scrapings (usually few or no oocysts, large number of merozoites). Differentiate from Duck viral hepatitis, Duck viral enteritis, anatipestifer.

Treatment

Sulphonamides (e.g. Sulphadimidine 30–600gm/100 birds/day, 3 days on, 2 days off, 3 days on), Amprolium, Vitamins A and K in feed or water.

Prevention

If required coccidiostats could be used in feed, however this is not routinely practised. Hygiene.

DUCK VIRAL HEPATITIS

Introduction

A viral disease of ducks occurring worldwide and previously a scheduled disease in UK. Morbidity is around 100% and mortality 0–95%. The disease is transmitted by infected ducks and other waterfowl and spreads rapidly, recovered birds carrying the virus for 8 weeks. The infective agent, a picornavirus may also survive for ten weeks in brooders and five weeks in faeces. A different picornavirus causes a similar condition in North America.

Signs

Sudden death.
Death in good condition.
Depression.
Fall on side, paddling of legs, arching of back, rapid deterioration and death, often in opisthotonus.

Post-mortem lesions

Liver swollen.
Punctate/diffuse haemorrhages.
Kidneys and spleen swollen.
Microscopically – focal necrosis, bile duct proliferation and inflammation.

Diagnosis

History, lesions, SN serology, isolation in CE (causes stunting of 9 day embryo). Differentiate from Duck plague (viral enteritis), Duck septicaemia (anatipestifer), coccidiosis, Newcastle disease, Influenza and a 'Type II Variant' hepatitis caused by Astrovirus.

Treatment
Antiserum, 0.5 ml serum of recovered birds given intramuscularly.

Prevention
Vaccination and/or antiserum, breeder vaccination. Live, only slightly attenuated vaccine is applied at day old by foot web stab and may be repeated in breeding birds to provide maternal immunity.

DUCK VIRUS ENTERITIS, DUCK PLAGUE

Introduction
A herpesvirus infection of ducks and geese diagnosed in the UK in 1972, mostly in ornamental collections, in USA since 1967, also the Netherlands and other countries. All waterfowl are susceptible and the Barbary duck is more susceptible than the Pekin. The disease follows a very acute course with a morbidity of 5–100% and mortality of 5–100%. Transmission is by infected birds, fomites and arthropods. Recovered birds may carry the virus for a year.

Signs
Sudden deaths.
Rapidly spreading disease.
Drop egg production.
Photophobia.
Ataxia.
Closed eyes.
Thirst.
Severe diarrhoea, sometimes dysentery.
Dehydration.
Paresis.
Tremor.
Occasionally penile prolapse in the penis in drakes.
Occasionally cyanosis of the bill in the young.

Post-mortem lesions
Severe enteritis.
Crusty plaques from oesophagus to bursa (covered by yellowish

plaques in later stages).
Haemorrhage in intestine, body cavities, heart, pericardium, liver, spleen.
Young ducks may show thymic and bursal lesions.

Diagnosis

Isolation: Duck CAMs 12 day embryos die in 4 days, HA-, intranuclear inclusions Differentiate from Duck hepatitis, oesophagitis (birds on restricted feed), vent gleet, pasteurellosis, coccidiosis.

Treatment

None, but vaccination in face of outbreak is of value, probably through interference.

Prevention

Isolation from waterfowl, vaccination if approved by authorities (CE adapted live virus).

Gizzard worms – Geese

Introduction

A nematode worm parasite, *Amidostomum anseris*, affecting geese and ducks. Worms develope to L3 in eggs and infection is by the oral route direct from environment.

Signs

Depression.
Loss in condition and weight.
Slow growth.

Post-mortem lesions

Ulceration, necrosis and partial sloughing of gizzard lining, muscular wall may be sacculated or ruptured.
Adults are 2–4 cm long and usually bright red.

Diagnosis

Lesions, visualisation of worms.

Treatment

Levamisole, benzimidazoles.

Prevention
Rotation of ground on annual basis.

GOOSE PARVOVIRUS (DERZSY'S DISEASE)

Introduction
This is a highly contagious condition of geese and young Muscovy ducks. The younger the bird affected the more acute the condition and the higher the mortality. Losses are negligible in birds over 5 weeks of age. It is caused by a parvovirus distinct from chicken and mammalian parvoviruses. The amount of maternal antibody passed from the breeding birds will affect the severity and timing and severity of the condition in the young birds. Vertical transmission resulting in congenital infection may occur.

Signs
Prostration and death in acutely affected goslings.
Reduced feed intake.
Excessive water intake.
Swollen eyelids and eye and nasal discharge.
Profuse white diarrhoea.
Membrane covering tongue.
Loss of down.
Reddening of skin.

Post-mortem lesions
Pale myocardium.
Swelling and congestion of liver, spleen and pancreas.
Fibrinous pericarditis.
Fibrinous perihepatitis.
Ascites.

Diagnosis
Signs and lesions in birds of the appropriate age and species.

Treatment
No specific treatment. Antimicrobials may be of value in reducing the effects of secondary bacterial infections.

Prevention

Hatching and brooding geese from different parent flocks together should be avoided. Ideally flocks that have suffered the disease should not be used for breeding as they may become persistent excreters of the infection. Administration of immune serum has been shown to be effective but may require two doses (day old and around 3 weeks). The preferred approach is to immunise breeding birds with an attenuated live vaccine.

MYCOPLASMA IMMITANS INFECTION

Introduction

Infection with *Mycoplasma immitans* is seen in geese, ducks (France), and partridges (UK).

Signs

Swollen sinuses.

Post-mortem lesions

Sinusitis.

Diagnosis

Shows cross reaction in IFA to M.g. although DNA homology is only 40%.

Treatment

As for M.g.

Prevention

As for M.g.

PSEUDOTUBERCULOSIS

Introduction

An infection of ducks, other poultry, wild birds, rodents and man with the bacterium *Yersinia pseudotuberculosis*. The disease has a mortality of 5–75%.

Signs

Weakness.

Ruffled feathers.
Diarrhoea.
Sometimes sudden death.

Post-mortem lesions
Miliary lesions in a variety of organs but especially in the liver. In peracute disease the only lesion may be enteritis.

Diagnosis
Lesions, isolation. Differentiate from Duck viral enteritis, duck viral hepatitis, colibacillosis, coccidiosis, tuberculosis.

Treatment
Tetracyclines.

Prevention
Good husbandry and hygiene, adequate protection, 'hardening off', sulphonamides in feed.

STREPTOCOCCUS BOVIS SEPTICAEMIA

Introduction
An infection of ducks with the bacterium *Streptococcus bovis*. This condition is a cause of sudden deaths in ducklings of 2–3 weeks of age.

Signs
Sudden deaths.
Leg weakness or incoordination in a few birds. Affected well-grown birds may appear to have died from smothering.

Post-mortem lesions
Beak cyanosis.
Lung congestion and oedema.
Liver congestion.
Splenic hyperplasia.

Diagnosis
History, lesions, isolation. Sodium deficiency can appear very similar at about the same age – affected birds respond well to

saline injection and will actively seek out salt scattered in the litter.

Treatment
Amoxycillin.

Prevention
Good husbandry and hygiene, extra care in the immediate post-brooding period.

APPENDICES 9

APPENDIX 1

A GLOSSARY OF TECHNICAL TERMS, AND ABBREVIATIONS

Aetiology – the cause of a disease

Airsacculitis – inflammation of the air sacs

Ataxia – loss of balance, staggering

Atrophy – reduction in size (of an organ)

Amorphous – lacking defined internal structure

Autogenous – made from material derived from the affected animal, flock or farm

Bacterins – vaccines made of bacteria which have been inactivated or killed

CAM – Chorio-alantoic membrane – part of the developing chick embryos used for the isolation of certain viruses (e.g. pox viruses)

Caseous – cheese-like

CE – Chick embryo

Ciliostatic – the ability to stop movement of cilia – usually the tiny hair-like structures lining the wind pipe whose function is to move the mucus upwards

Coligranuloma – A chronic abscess-like condition associated with some *E. coli* infections

Conjunctivitis – inflammation of the membrane covering the insides of the eyelids and the front surface of the eyeball

Cuffing – forming a cuff

Cyanosis – blue to purple appearance of the tissues

Diarrhoea – abnormally liquid faeces

DID – Double immunodiffusion – a serological test which may be used either to detect an antibody response or to identify the presence of an antigen (e.g. in tissues)

Diuresis – excessive production of urine

Dyspnoea – difficulty in breathy

Dysentery – diarrhoea with blood

Ecchymoses – moderate to large haemorrhages in tissues

Elisa – Enzyme-linked Immuno-sorbent Assay – a test used mainly to detect and quantify antibody levels in blood

Emphysematous – presence of air where it shouldn't be

Eosinophilic – tissues selectively taking up the pigment eosin when stained which gives a reddish colouration

Epiphyseal – pertaining to the ends of the bones

Epiphyses – the ends of long bones

Erythroblastosis – an excessive growth of the erythroblasts – the progenitor cells that give rise to the erythrocytes

Erythrocytes – red blood cells

Exsanguination – loss of the majority of the circulating blood

Facultative – can either take it or not – e.g. a Facultative Anaerobe can grow either in the presence or absence of oxygen

FAT – fluorescent antibody test – usually used to demonstrate an antigen in tissues

Fibrinous – of or pertaining to fibrin – the protein which forms from circulating blood protein in a blood clot

Fibrosarcoma – a cancer of the cells that produce interstitial fibrous tissue

Filarial – a type of worm

Flexion – a bending of a joint

Fomites – inanimate objects that may carry and transmit infection

Granuloma (plural granulomata) – a focal lesion of chronic inflammation – often appears as a yellow spot or nodule.

HA+ – presence of a haemagglutinin – an agent that causes red cells to clump

HI – Haemagglutination Inhibition test – a test for antibody used for Newcastle disease, Infectious Bronchitis etc.

Hepatitis – inflammation of the liver

Hepatomegaly – enlargement of the liver

Hyphae – the fine hair like structures of fungi infecting tissues

IFA – indirect fluorescent antibody

Inappetance – reduction in feed intake

Intermediate host – some parasites must infect another animal species before completing their life cycle in the definitive host

Keratitis – inflammation of the cornea

Lentogenic – slow growing – applied to mild vaccinal strains of Newcastle disease virus

Lymphoblastic – pertaining to cells that are precursors of lymphocytes, the lymphoblasts

Megaschizonts – very large Schizonts – the asexual reproductive stage of coccidia

Meningitis – inflammation of the meninges, the membrane overing the brain and spinal cord

Mesogenic – moderately fast growing – applied to semi-pathogenic Newcastle disease viruses

Miliary – spread throughout the tissues

Mycotic – of or pertaining to fungi

Myelocytes – cells from the bone marrow, precursors of granulocytic blood cells

Myxosarcoma – malignant tumours of connective tissue

NAD – Nicotinamide Adenine Dinucleotide – a nutrient required for growth by certain bacteria

Necrosis – death of areas of tissue prior to the death of the animal

Ocular – pertaining to the eye

Opisthotonus – head pulled back in a 'star-gazing' posture

Opthalmitis – inflammation of the structures of the eye

Osteochondritis – inflammation of bone and cartilage

Osteopetrosis – a specific tumour-type disease causing malformation of bones

PAGE – polyacrylamide gel electrophoresis – a method of identifying proteins and some infectious agents

Papules – focal nodules or small plaques in skin

Paralysis – loss of control of skeletal muscles

Paratenic – a non-definitive host that allows parasite growth or partial maturation

Parenchymatous – refers to organs with functional tissue (parenchyma) not just support tissue, e.g. liver

Paresis – a milder form of paralysis

Pathognomonic – typical and diagnostic of the condition

PCR – Polymerase Chain Reaction – a method of dramatically replicating nucleic acids

Pericarditis – inflammation of the pericardium

Perihepatitis – inflammation of the capsule of the liver that results in the formation of a layer of fibrin or pus on the surface of the liver – see figure 25.

Peritonitis – inflammation of the peritoneum – the membrane covering abdominal organs

Perivascular – around blood vessles

Petechiae – tiny haemorrhages in tissues

Prepatent period – the interval between infection and the start of egg laying by a parasite

Purulent – of or pertaining to pus, an accumulation of dead inflammatory cells

Recumbency – lying down

SAG – serum agglutination – a type of serological or blood test

Schizonts – the asexual reproductive stage of coccidia

Sinusitis – inflammation of the sinuses

Splenomegaly – enlargement of the spleen

Synovitis – inflammation of the synoviae, the lining of joints

Torticollis – twisted neck

Transport host – another animal species that harbours an infection and physically transports it to a new host. It is not required to complete the life cycle, and the infection does not develop in it.

Valgus and Varus – a defect of leg bones resulting in the leg twisting outward or inward

Vasocontriction – constriction of blood vessels

Velogenic – fast growing – applied to virulent strains of Newcastle diseae virus

Vesicles – blisters in skin or mucous membrane with fluid

Viscera – the internal organs – typically refers to intestinal tract, liver, kidney, spleen etc.

Viscerotropic – with a tendency to affect the viscera, though not necessarily all viscera

APPENDIX 2

ONE ANSWER TO THE TASK A

Having regard for biosecurity objectives take the case of a 15 000 bird broiler parent laying site, review the following list of activities and place them in a chronological order. What would you do first, second, third etc? The history for this particular flock was that a positive *Salmonella* Enteritidis isolate had been obtained from the hatchery during the week of depleting the previous flock. Government authorities declined to take carcases to establish their true infection status. The list of activities is currently totally random. Be prepared to explain your reasoning. Feel free to add other activities to the list. It might be appropriate to include some activities more than once in your list.

ACTIVITY	ORDER	WHY?
Blood-sample birds for S.E.	1.	Try to establish true status
Cut weeds	2.	Reduce rodent cover around
Remove birds	3.	Remove infected birds
Spray house with insecticide	4.	Control beetles
*Blow down dust	5.	To remove litter
Remove litter	6.	Remove bulk of infected material
Re-wire fans (Health & Safety)	7.	Get on with maintenance
Wash house	8.	Remove as much organic material as possible
Wash nests and slats	9.	Ditto – concrete pads are helpful
Clean and disinfect staff room	10.	May also be contaminated
Fix broken wall boards	11.	Provide good surfaces to disinfect
Cement up cracks in dwarf walls	12.	Ditto
Set up equipment	13.	Put clean equipment in (may disinfect first)
Disinfect house	14.	Decontaminate building and equipment

Poultry Diseases Pocket Guide

ACTIVITY	ORDER	WHY?
Install foot-baths	15.	Full barrier as soon as house disinfected
Close up cracks under doors	16.	Avoid rodent recolonisation, gas proof
Clean and disinfect bait boxes	17.	Contaminated material
Order shavings	18.	Should not arrive prior to disinfection
Order feed	19.	Ensure feed system cleaned, perhaps sanitised
Remove debris around houses	20.	Avoid rodent cover
Inspect house	21.	Have we missed anything? Redisinfect?
*Hygiene check/Salm.	22.	Ideally with time to redisinfect if poor/positive
Sample litter for *Salmonella*	23.	Reject if positive – monitor supplier at source?
Fumigate house	24.	Ideally formalin for 24 hours + then ventilate
Staff holiday	25.	When rest done! Useful while fumigating!
New or laundered overalls in use	26	Back to normal operations
Re-bait rodent boxes	27.	Maintain effective control throughout
Receive pullets	28.	Back in action.

* indicates steps not in original list.

APPENDIX 3

SOME USEFUL TABLES

Units of Measurement used in Biological Science

There are two methods used; - the imperial system, also called the Foot Pound Second and the metric system, based upon the Centimetre Gram and Second. The metric system was modernised in 1960 and an international system of units (SI) agreed. This is now used throughout the world.

Most poultry-producing countries use the metric system but in some, part metric and part imperial are also used, for example in the USA.

This book uses the metric system throughout the text but for ease of reference, some tables include imperial measurements.

METRIC UNITS AND RELATIVE VALUES

Weight	Volume	Length	Relationship
Gram (gm)	Litre (l)	Metre (m)	1
Kilogram (kg)	Kilolitre (kl)	Kilometre (km)	1,000
Decigram (dg)	Decilitre (dl)	Decimetre (dm)	1/10
Centigram (cg)	Centilitre (cl)	Centimetre (cm)	1/100
Milligram (mg)	Millilitre (ml)	Millimetre (mm)	1/1,000
Microgram (µg or mcg)	Microlitre (µl)	Micrometre (µm)	1/1,000,000
Nanogram (ng)	Nanolitre (nl)	Nanometre (nm)	1/1,000,000,000
Picogram (pg)	Picolitre (pl)	Picometre (pm)	1/1,000,000,000,000

( ) = abbreviations

Poultry Diseases Pocket Guide

HOW TO CONVERT UNITS OF MEASUREMENTS

Imperial to Metric

To convert	Into	multiply by
Length		
Inches	millimetres	25.4
Inches	centimetres	2.54
Feet	Metres	0.3048
Yards	Metres	0.9144
Miles	kilometres	1.6093
Area		
square inches	square centimetres	6.4516
square feet	square metres	0.093
square yards	square metres	0.836
Acres	Hectares	0.405
square miles	square kilometres	1.6093
Volume		
cubic inches	cubic centimetres	16.387
cubic feet	cubic metres	0.0283
cubic yards	cubic metres	0.7646
fluid ounces	Millilitres	28.41
Pints	Litres	0.568
Gallons	Litres	4.55
Weight		
Ounces	Grams	28.35
Pounds	Kilograms	0.45359
Tons	Kilograms	1016.00
Tons	Tonnes	1.016

HOW TO CONVERT UNITS OF MEASUREMENTS

Metric to Imperial

To convert	Into	multiply by
Length		
Millimetres	inches	0.0394
Centimetres	inches	0.3937
Metres	feet	3.2808
Metres	yards	1.0936
Kilometres	miles	0.6214
Area		
square centimetres	square inches	0.155
square metres	square feet	10.764
square metres	square yards	1.196
hectares	acres	2.471
square kilometres	square miles	0.386
Volume		
cubic centimetres	cubic inches	0.061
cubic metres	cubic feet	35.315
cubic metres	cubic yards	1.308
litres	pints	1.760
litres	gallons	0.220
Weight		
grams	ounces	0.0352
kilograms	pounds	2.2046
kilograms	tons	0.000984
tonnes	tons	0.9842

Poultry Diseases Pocket Guide

Quick Conversion Tables

LENGTH - Inches - Millimetres - Centimetres

Approximate 1 in = 25 mm (25.4 mm) 1 mm = $1/_{32}$ in (0.039 in)
= 2.5 cm (2.54cm) 1 cm = $2/_{5}$ in (0.393 in)
(Exact values)

Inches		Millimetres	Inches		Centimetres
	$1/_4$	6.4		$1/_4$	0.6
	$1/_2$	12.7		$1/_2$	1.3
	$3/_4$	19.0		$3/_4$	1.9
0.04	**1**	25.4	0.39	**1**	2.5
0.08	**2**	50.8	0.79	**2**	5.1
0.12	**3**	76.2	1.18	**3**	7.6
0.16	**4**	101.6	1.57	**4**	10.2
0.20	**5**	127.0	1.97	**5**	12.7
0.24	**6**	152.4	2.36	**6**	15.2
0.28	**7**	177.8	2.76	**7**	17.8
0.31	**8**	203.2	3.15	**8**	20.3
0.35	**9**	228.6	3.54	**9**	22.9
0.39	**10**	254.0	3.94	**10**	25.4
0.43	**11**	279.4	4.34	**11**	27.9
0.47	**12**	304.8	4.72	**12**	30.5

Chapter 9 - Appendices

LENGTH - Feet - Yards – Metres

| 1ft = 0.3 m (0.3048) | 1m = 3ft $^3/_8$in (3.2808 ft) |
| 1yd = 0.9m (0.9144) | 1m = 1yd 3in (1.0936 yds) |

Feet		Metres	Yard		Metres
3.3	1	0.3	1.1	1	0.9
6.6	2	0.6	2.2	2	1.8
9.8	3	0.9	3.3	3	2.7
13.1	4	1.2	4.4	4	3.7
16.4	5	1.5	5.5	5	4.6
19.7	6	1.8	6.6	6	5.5
23.0	7	2.1	7.7	7	6.4
26.2	8	2.4	8.7	8	7.3
29.5	9	2.7	9.8	9	8.2
32.8	10	3.0	10.9	10	9.1
49.2	15	4.6	16.4	15	13.7
65.6	20	6.1	21.9	20	18.3
82.0	25	7.6	27.3	25	22.9
98.4	30	9.1	32.8	30	27.4
114.8	35	10.7	38.3	35	32.0
131.2	40	12.2	43.7	40	36.6
147.6	45	13.7	49.2	45	41.1
164.0	50	15.2	54.7	50	45.7

Poultry Diseases Pocket Guide

LENGTH - Miles - Kilometres

1 mile = 1.6 km (1.6093km)
1 km = $^5/_8$ mile (0.6214 mile)

Miles		Kilometres
0.6	**1**	1.6
1.2	**2**	3.2
1.9	**3**	4.8
2.5	**4**	6.4
3.1	**5**	8.0
3.7	**6**	9.7
4.3	**7**	11.3
5.0	**8**	12.9
5.6	**9**	14.5
6.2	**10**	16.1
12.4	**20**	32.2
18.6	**30**	48.3
24.9	**40**	64.4
31.1	**50**	80.5

AREA
Square feet - Square metres

1 sq ft = 0.1 m^2 (0.0929 m^2)
1m^2 = 10$^3/_4$ ft^2 (10.764 sq ft)

Square Feet		Square Metres
10.8	**1**	0.09
21.5	**2**	0.19
32.3	**3**	0.28
43.1	**4**	0.37
53.8	**5**	0.46
64.6	**6**	0.56
75.3	**7**	0.65
86.1	**8**	0.74
96.9	**9**	0.84
107.6	**10**	0.93
118.4	**11**	1.02
129.2	**12**	1.11
139.9	**13**	1.21
150.7	**14**	1.30
161.5	**15**	1.39
172.2	**16**	1.49
183.0	**17**	1.58
193.8	**18**	1.67
204.5	**19**	1.77
215.3	**20**	1.86

WEIGHT
Pounds - Kilograms

1 pound - 0.5 kg (0.454 kgs)
1 kg = 2$^1/_4$ lbs (2.205 lbs)

Pounds		Kilograms
2.2	**1**	0.45
4.4	**2**	0.91
6.6	**3**	1.36
8.8	**4**	1.81
11.0	**5**	2.27
13.2	**6**	2.72
15.4	**7**	3.18
17.6	**8**	3.63
19.8	**9**	4.08
22.2	**10**	4.54
24.3	**11**	4.99
26.5	**12**	5.44
28.7	**13**	5.90
30.9	**14**	6.35
33.1	**15**	6.80
35.3	**16**	7.26
37.5	**17**	7.71
39.7	**18**	8.16
41.9	**19**	8.62
44.1	**20**	9.07

Poultry Diseases Pocket Guide

VOLUME - Imperial / Metric

1 imperial pint = 0.5 litre (0.568 litres)
1 litre = $1^3/_4$ pints (1.7598 imp pints) = $1/_4$ gallon (0.22 gallon)
1 gallon (8 pints) = 4.5 litres (4.546 litres)

Imp. Pints		Litres	Imp. Gallons		Litres
1.76	**1**	0.57	0.22	**1**	4.55
3.52	**2**	1.14	0.44	**2**	9.09
5.28	**3**	1.70	0.66	**3**	13.64
7.04	**4**	2.27	0.88	**4**	18.18
8.80	**5**	2.84	1.10	**5**	22.73
10.56	**6**	3.41	1.32	**6**	27.28
12.32	**7**	3.98	1.54	**7**	31.82
14.08	**8**	4.55	1.76	**8**	36.37

Note:
One American Pint = 0.75 Imperial Pints
One American Gallon = 0.75 Imperial Gallon
One American Gallon = 3.785 Litres

TEMPERATURE CONVERSION °C - °F

$°F = (°C \times 1.8) + 32 \qquad °C = (°F - 32) \div 1.8$

°C	⇌	°F	°C	⇌	°F	°C	⇌	°F
-27		-16.6	1		33.8	29		84.2
-26		-14.8	2		35.6	30		86.0
-25		-13.0	3		37.4	31		87.8
-24		-11.2	4		39.2	32		89.6
-23		-9.4	5		41.0	33		91.4
-22		-7.6	6		42.8	34		93.2
-21		-5.8	7		44.6	35		95.0
-20		-4.0	8		46.4	36		96.8
-19		-2.2	9		48.2	37		98.6
-18		-0.4	10		50.0	38		100.4
-17		+1.4	11		51.8	39		102.2
-16		3.2	12		53.6	40		104.0
-15		5.0	13		55.4	41		105.8
-14		6.8	14		57.2	42		107.6
-13		8.6	15		59.0	43		109.4
-12		10.4	16		60.8	44		111.2
-11		12.2	17		62.6	45		113.0
-10		14.0	18		64.4	46		114.8
-9		15.8	19		66.2	47		116.6
-8		17.6	20		68.0	48		118.4
-7		19.4	21		69.8	49		120.2
-6		21.2	22		71.6	50		122.0
-5		23.0	23		73.4	51		123.8
-4		24.8	24		75.2	52		125.6
-3		26.6	25		77.0	53		127.4
-2		28.4	26		78.8	54		129.2
-1		30.2	27		80.6	55		131.0
0		32.0	28		82.4	56		132.8

INDEX

A

Adenovirus, 118, 136,138,174,220
Aflatoxins, 157
Amyloidosis, 73
Anatipestifer disease, 241
Angara disease,136
Annexe 4, 52
Antibiotics, 42–53
Antibodies, 32
Antimicrobial medication,
    see Medicines, antimicrobial
Aortic rupture, 219
Aplastic anaemia, 135
Argas persicus, *192, 198,*
Arizona hinshawii, 215
Arizona infection, 215
Arizonosis, 215
Arthritis, staphyloccocal, 195
Arthritis, viral, 204
Ascaridia, 180
Ascites, 74
Aspergillosis, 75
Aspergillus flavus, 157
Aspergillus fumigatus, 75
Aspergillus ochraceus, 157
Avian encephalomyelitis,
    egg drop, 76
Avian encephalomyelitis,
    tremors, 77
Avian infectious hepatitis, 203
Avian influenza-highly
    pathogenic, 127
Avian leukosis, 79
Avian leukosis, sero-type J, 86
Avian monocytosis, 172
Avian pox, 130
Avian rhinotracheitis, 82

B

Baby chick nephropathy, 205
Bacillary white diarrhoea, 184
Bacterial chondronecrosis with
osteomyelitis, 124
Beak necrosis, 84
Bedbug infestation, 84
Big liver and spleen disease, 85
Biosecurity procedures, disadvantages
    of, 24–25
Biosecurity,
    definition, 11
    principles of, 11–27
Biotin deficiency, 86, 99 195, 208
Birnavirus, 144
Biting lice, 87
Blackfly infestation, 88
Blackhead, 222
Bluecomb, chickens, 172
Bluecomb, turkeys, 233
Bordetella avium, 234
Borrelia anserina, 192
Botulism, 89
Brachyspira pilosicoli, 192
Breast blister, 90
Broiler ascites syndrome, 74
Bumble foot, 195
Bursa of Fabricius, 31

C

Caecal worm, 91
Cage fatigue, 166
Calcium deficiency, 166
Calcium tetany, 92
Campylobacter infection, 92
Campylobacter jejuni, 92
Candida albicans, 94
Candidiasis, 94
Canker, 199
Cannibalism, 95
Capillaria, 96
Capillariasis, 96, 112
Carcases, disposal, see dead birds,
    disposal
Cellulitis, 97
Cestodes, 197
Chicken anaemia, 98
Chicks,
    antimicrobial medication, 46
    as a major input, 19
Chlamydia psittaci, 216
Chlamydiosis, 216
Chlorination of water, 22–23

Chondrodystrophy, 99
Chronic respiratory disease
  chickens, 153
  turkeys, 226
Cleaning and disinfecting houses and equipment, 18–19
Clostridium botulinum, 89
Clostridium colinum, 202
Clostridium perfringens, 159
Clostridium septicum, 132
Coccidiosis,
  caecal, *E. tenella*, 106
  ducks, 243
  E. mitis, 104
  E. praecox, 105
  geese, 243
  ileorectal, *E. brunetti*, 108
  intestinal, of ducks and geese, 243
  kidney, 242
  mid-intestinal, *E. maxima*, 102
  mid-intestinal, *E. necatrix*, 103
  turkeys, 217
  upper intestinal, *E. acervulina*, 100
Colibacillosis, 109
Colisepticemia, 109
Contact dermatitis, 111
Copper deficiency, 219
Coronavirus, 140, 142, 143, 144, 233
Crop, impaction, see impaction and foreign bodies of gizzard
Crop, pendulous, 170
Cropworms, 112
Cryptosporidium baileyi, 113
Cryptosporidium meleagridis, 113
Cryptosporidiosis, 113

D

Dactylaria gallopava, 114
Dactylariosis, 114
Dead birds, disposal, 24
Deep pectoral myopathy, 162
Degenerative joint disease, 115
Depluming mites, 116
Dermanyssus gallinae, 173
Derzsy's disease, 247
Dissecting aneurysm, 219

Dosage levels, understanding, 40
Drinking water medication, advantages and disadvantages, 46
Drinking water,
  administering medicine, 41–43
  as a major input, 22–23
  hygiene, 22–23
Duck plague, 245
Duck viral hepatitis, 244
Duck virus enteritis, 245
Dysbacteriosis, 116

E

Eastern equine encephalomyelitis, 121
Egg drop, avian encephalomyelitis, 76
Egg drop syndrome-76, 118
Eimeria spp, see entries under coccidiosis
Electrolytes, 45
Encephalomalacia, 210
Encephalomyelitis, avian, 76
Encephalomyelitis, equine, 121
Endocarditis, 119
Enterovirus, 150
Epidemic tremors, 77
Epiphysiolysis, 120
Equine encephalomyelitis, 121
Erysipelas, 121
Erysipelothrix insidiosa, 121
Exudative diathesis, 210

F

Fatty liver and kidney syndrome, 86
Fatty liver haemorrhagic syndrome, 123
Favus, 123
Feather pecking, 95
Feed
  additives, 52–56
  hygiene, 20–22
  medication, 52–56
Femoral head necrosis, 124
FHN, 124
Flipover, 196
Foreign bodies of gizzard, 137

Index

Fowl cholera, 125
Fowl plague, 127
Fowl pox, 130
Fowl typhoid, 182
Frounce, 199

G

Gangrenous dermatitis, 132
Gape, 133
Gizzard worms,
  chickens, 134
  geese, 246
Gizzard, impaction and foreign bodies, 137
Gongylonema ingluvicola, 112
Goose parvovirus, 247
Gumboro, 144

H

HACCP, 15
Haemorrhagic anaemia, 135
Haemorrhagic disease, 135
Haemorrhagic enteritis, 220
Hairworm infection, 96
Health plans, veterinary, 33, 56–59
Heat stress, 134
Herpesvirus, 148, 151, 245
Heterakis gallinae, 91, 222
Hexamita meleagridis, 221
Hexamitiasis, 221
Histomonas melagridis, 222
Histomoniasis, 222
Histomonosis, 222
Hock burn, 111
HPAI, 127
Hydropericardium-hepatitis syndrome, 136

I

IB, see Infectious bronchitis
IBD, 144
ILT, 148
Immune system, 31–33
Immunisation, 32–33
Immunisation programmes, development of, 33
Immunity,
  active, 32
  antibody, 31
  cellular, 31
  passive, 32
Impaction of gizzard, 137
Inclusion body hepatitis, 138
Individual birds, treating, 44–45
Infectious bronchitis, 140
  793b variant sudden death syndrome in broiler parents, 142
  egg-layers, 143
Infectious bursal disease, 144
Infectious coryza, 147
Infectious laryngotracheitis, 148
Infectious sinusitis, 226
Infectious synovitis, 155
Influenza viruses, 127
Intussusception, 149

J

Joint disease, degenerative, 115

K

Kinky-back, 193
Knemidocoptes, *116*

L

Leucocytozoon species, 224
Leucocytozoonosis, 224
Leukosis,
  avian lymphoid, 79
  avian serotype/subtype J, 86
  myelocytoma, 86
Leukosis/sarcoma group, 79
Litter, as a major input, 23–24
LPD, 225
Lymphoid leukosis, 79
Lymphoid tumour disease, 176
Lymphoproliferative disease, 225

Poultry Diseases Pocket Guide

M

Malabsorption syndrome, 150
Management, poultry health, 5–7
Marek's disease, 151
Market demands, satisfying, 4
Medication, 37–59
   oral, 41–44
   water, 41–43
Medicines,
   antimicrobial, 45–46
   controlling and storing, 50–51
   disposing of, 51–52
   in feed, 43–44
   legal requirements, 38–40
   orally, 41–44
   prescription of, 40
   recording of use, 48–50
   responsible use of, 47–48
Menocanthus stramineus,
M.g., see *Mycoplasma gallisepticum* infection
M.i., 227
Mild respiratory disease, 174
Mites,
   depluming, 116
   northern fowl, 173
   red, 173
   scaly leg, 116
M.m., 228
Monilia, 94
Moniliasis, 94
M.s., 155
Muscular dystrophy, 210
Mycoplasma gallisepticum infection,
   chickens, 153
   turkeys, 226
Mycoplasma immitans infection, 248
Mycoplasma iowae infection, 227
Mycoplasma meleagridis infection, 228
Mycoplasma synoviae infection, 155
Mycoplasmosis, see separate entries under *Mycoplasma* spp
Mycotoxicosis, 157
Myelocytomatosis, 86

N

Necrotic dermatitis, 132
Necrotic enteritis, 159
Nephrosis, 205
New duck syndrome, 241
Newcastle disease, 167
Nicarbazin, 54–56
Non-specific bacterial enteritis, 116
Non-starters, 161
Northern fowl mite, 173
Nutritional roup, 207

O

Omphalitis, 211
Oregon disease, 162
Ornithobacterium infection, 163
Ornithobacterium rhinotracheale, 163
Ornithosis, 216
ORT, 163
Orthomyxovirus, 127
Osteomyelitis/chondritis, 124
Osteomyelitis complex, turkeys, 230
Osteoporosis, 166

P

Paracolon infection, 215
Paramyxovirus 1 (PMV-1), 167
Paramyxovirus 2 (PMV-2), 170
Paramyxovirus 3 (PMV-3), 230
Paramyxovirus 6 (PMV-6), 231
Paratyphoid infections, 185
Parvovirus, goose, 247
Pasteurella multocida, *125*
Pasteurellosis, 125
PEMS, 232
Pendulous crop, 170
Perosis, 99
Picornavirus, 244
Pneumovirus, 82, 174, 235, 236
Pododermatitis, 111
Pox, 130
Proventricular worms, 171

Proximal femoral degeneration, 124
Pseudotuberculosis, 248
Psittacosis, 216
Pullet disease, 172
Pullorum disease, 184

Q

Quail disease, 202

R

Recording, medicines use, 48–50
Red mite, 173
Reovirus, 150, 195, 204
Residues,
   avoidance, 4, 52
   nicarbazin, 54–56
   maximum limit, 52
Respiratory adenovirus infection, 174
Respiratory disease complex, 174
Reticuloendotheliosis, 176
Retrovirus, 79, 176, 225
Rickets, hypocalcaemic, 177
Rickets , hypophosphataemic, 178
Riemerella anatipestifer, 241
Rotavirus infection, 150, 179
Roundworm, 180
Runting, 150
Ruptured gastrocnemius tendon, 181

S

Salmonella Arizonae, 215
Salmonella Gallinarum, 182
Salmonella Pullorum, 184
Salmonellosis, 185
   S. Enteritidis and *S.* Typhimurium infections, 187
Salpingitis, 190
Scaly leg mites, 116
Self inoculation, 41
Shaky leg syndrome, 233
Slipped tendon, 99
Spiking mortality,
   of chickens, 191
   of turkeys, 232
Spirochaetosis, 192
Splay leg, 194
Spondylolisthesis, 193
Spraddle legs, 194
Staphylococcal arthritis, 195
Staphylococcosis, 195
Starve-outs, 161
Streptococcus bovis septicaemia, 249
Stunting, 150
Sudden death syndrome, 196
Sudden death syndrome, broiler parents, 142
Swollen head syndrome, 82
Syngamus trachea, 133

T

Tapeworms, 197
TD, 198
Tetratrichomonas, 221
Thrush, 94
Tibial dyschondroplasia, 198
Ticks, 198
Transmissible enteritis, 233
Trichomoniasis, 199
Tuberculosis, 200
Turkey coryza, 234
Turkey rhinotracheitis,
   adult, 236
   in rear, 235
Turkey viral hepatitis, 237
Twisted leg, 201

U

Ulcerative enteritis, 202

V

Vaccines, how they work, 31–33
Venezuelan equine encephalomyelitis, 121
Veterinary health plans, 33, 56–59
Vibrionic hepatitis, 203
Viral arthritis, 204

Poultry Diseases Pocket Guide

Visceral gout, 205
Vitamin A deficiency, 207
Vitamin B deficiencies, 208
Vitamin E deficiency, 210

W

Water medication, 41–43
Water hygiene, 22–23

Water-belly, 74
Western equine encephalomyelitis, 121

Y

Yolk sac infection, 211
Yersinia pseudotuberculosis, *248*
Yucaipa disease, 170